AT HOME IN THE SUN

An Open-House Tour
of Solar Homes in the United States

by

Norah Deakin Davis
and Linda Lindsey

GARDEN WAY PUBLISHING
Charlotte, Vermont 05445

Library of Congress Cataloging in Publication Data

Davis, Norah Deakin, 1941–
   At home in the sun.

   Bibliography: p.
      1. Solar houses—United States.
I. Lindsey, Linda, 1942–   joint author.
II. Title.
TH7414.D39   728.6'9   79-12867
ISBN 0-88266-152-3
ISBN 0-88266-151-5 pbk.

To Our Fathers

Contents

9

29

36

42

56

63

90

104

154

166

183

198

211

223

Foreword

For some three billion years all the earth's creatures have been at home in the sun, and we expect this trend to continue. Human technical ingenuity finds its noblest outlet in building for a future when we will be, as Richard Brautigan put it, all watched over by machines of loving grace. In a soft energy future, the energy systems we build for warmth and cooling, transport, industry, and agriculture would improve our lives without taking their toll upon personal independence, environmental health, genetic diversity, or sustainability.

At Home in the Sun is an appropriate book for an appropriate technology. The decentralized solar movement needs books like this to keep in touch with itself. Attitudes toward convenience and concern for the earth as a living system are discussed, as they are integral to the reasoning behind solar energy utilization and the conservation of nonrenewable fuels. A careful look is taken, objectively and subjectively, at the failures and successes of the recent pioneers in the modern solar age. This open-minded approach is indispensable, for if we don't learn about our mistakes as well as our victories, the emergence of our most promising post-oil energy source will be slowed. Delay will inevitably bring inflation and the diversion of our limited capital resources to less desirable energy alternatives, making solar ever more difficult (though no less essential) to achieve.

We in Friends of the Earth applaud all who now give their time, creativity, and resources to further a solar future, and are grateful to the authors and publishers of *At Home*. We look forward to the time when all people and all their institutions will think solar. The sun got us this far, and stands the best chance of going the distance.

David R. Brower
President, Friends of the Earth
January 10, 1979

Preface

These words are being written in a room heated by the sun. The clacking of typewriter keys is mingling with the steady hum of a solar collector pump. Both of us—the authors of this book—are solar homeowners. Between us, we have five years of firsthand experience in living with the sun. We know that solar heating works and makes economic sense if you do it right.

Having at one time been innocents in solar toyland, however, we're well aware that difficulties await the neophyte who goes into it with too little preparation. At this stage in its development solar energy is in need of a realistic appraisal. We want a candid look at problems as well as benefits.

Those who brush solar homes with rose-colored paint are doing the cause as much harm as those who sabotage it with gloomy pessimism. The utility company executive who claims that solar heating will not be here for another decade creates a self-fulfilling prophecy. On the other hand, those who encourage the public to expect too much too soon run the risk of a backlash resulting from solar failures. The Joneses, finding their collector is beset with difficulties or discovering their investment won't be repaid until they're senior citizens, may tell their neighbors to forget it.

In selecting houses to include in this book, we were tempted to steer clear of that kind and pick only successful examples. Instead we made a point of asking owners about their mistakes in the hopes that those who build in the future can avoid the pitfalls. We tried to write the book we wish we'd had before we designed our own homes. We made selections partly on the basis of esthetic qualities and economic feasibility, but also because of the owners' willingness to pass along the benefits of their experience.

A delicate balance exists between being honest about problems and frightening people away from solar heating. We have no desire to magnify the difficulties out of proportion. Getting the bugs out of a new technology takes time, but as solar heating grows out of its infancy into childhood, leaving behind the era of one-of-a-kind systems, the bugs are gradually being exterminated.

In order to look at solar homes that have been lived in for some time, we selected a couple from the earliest experimental phase and a number from the next generation built from the time of the energy crisis in 1974 up to about 1977. The rest are a sample from the third generation constructed between 1977 and 1978, consisting largely of the latest in passive designs.

Like all technical subjects, solar energy is in danger of drowning in its own jargon. Words like "retrofit" are almost indispensable, but in a nontechnical book like this we can get by without brutes like "Delta T." A table is included for each house to satisfy those who crave technical data and to facilitate quick comparisons. Any item missing for a given house is either inapplicable or something the owners prefer not to publish. Total building costs, where printed, exclude the price of the land but include the cost of solar equipment. Information on operating and fuel costs may be imprecise—understandably so, since it's difficult for owners to determine, for example, how much electricity is consumed by the machinery and controls for a collector. In most cases the systems have not yet incurred maintenance costs.

We wish to thank the solar homeowners for their hospitality and regret that many had to be left out because of space limitations. Thanks are also due to Randy Scoog of Ecotope Group and Rebecca Voories of the Colorado Solar Energy Society for their help in locating solar homes, to Stephen B. Papazidis and Claude Bolduc for their photographic work, to Dolores Jordan and Penny Grover for typing assistance, to Joseph Holmes and Stephen O. Andersen for technical assistance, to Harris Hyman and William Shurcliff for reading the manuscript, and most of all to Steve Wolcott and Dick Davis for everything else.

Introduction

Since the time when we built our homes, solar heating has indeed been growing up. The Department of Energy (DOE) estimates that the United States now has 40,000 solar installations compared to a mere twenty-four in 1970. This exponential growth means that by 1985 conceivably one out of every ten Americans may live in a solar home or a home with solar-heated hot water. Predictions vary from 1.3 million solar buildings by Energy Secretary James Schlesinger to 11 million by the Solar Energy Industries Association.

No doubt many Americans are taking a wait-and-see attitude. They're waiting for expected cost reductions from the mass production of solar equipment. Manufacturers, however, are reluctant to invest the kind of money needed for tooling up until sufficient demand exists. Even when they come, assembly lines may not significantly lower prices, because of inflationary increases in the cost of labor, transportation, and particularly materials, which are mostly mass produced already and not apt to go down. One solar company claims it has made the manufacturing process as efficient as it's going to get, having reduced labor to less than one-tenth of production costs. Using cheaper materials might help, but simpler systems may be the only sure solution.

A few individuals may be afraid to invest in a solar collector on the grounds that a better design may come along tomorrow. But there is no way a system can become obsolete when it successfully heats a home with minimal operating and maintenance costs. How can you do better? No system is outdated if it does its job well.

The Basics of Solar Heating

To understand solar heating it will be helpful to look at the principles of radiation, conduction, and convection. The sun and other heat sources such as wood stoves transmit a form of radiant energy. Solar radiation reaches the earth in a broad spectrum of wavelengths. In solar heating we are concerned with a small portion of the spectrum including visible light and the immediately adjacent wavelengths, ultraviolet on one end of the spectrum and infrared on the other. Ultraviolet is largely filtered out of our atmosphere by the ozone layer, but enough reaches the earth to cause sunburn and skin cancer. Infrared radiation, whether from the sun or a hot object, is felt as heat.

In itself solar **radiation** is not hot; "heat" does not exist until the sun's rays strike a surface. A sunbeam passes through millions of miles of space without heating anything. When it finally strikes an object, it is either reflected or absorbed depending on the nature of the object. When absorbed it excites the molecules that compose the object, creating heat. White and shiny metallic surfaces reflect nearly all the light while objects of a particular color reflect only that color. Black surfaces absorb most of the radiation and are therefore used in solar collectors.

Transfer of heat by **conduction** is the direct transmittance of energy from molecule to molecule. Materials differ in their abilities to conduct heat. Metal and stone, being dense, are good conductors because their molecules are tightly packed and more easily excite one another. If you heat one end of a metal rod, the heat will be quickly conducted to the other end. If you then grasp the rod, your hand will heat up too, by conduction.

Air is a poor conductor. Because air molecules are spread out, they do not easily come into contact with each other, making air a fairly good insulator when contained to restrict its flow. When air molecules are unrestrained and agitated by contact with a heat source, they spread out further; this less dense, warm air will rise as long as denser, cool air is available to replace it. As the cooler air

1

Radiation

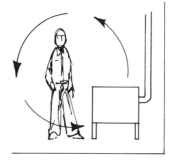

Convection

Conduction

moves into contact with the heat source, it in turn is heated and rises. This process is known as **convection.** It occurs in all fluids, both liquids and gases.

Evolution of Home Heating

We humans live in an envelope of air. To maintain our body temperature at 98.6° we need to prevent excessive radiative, conductive, and convective heat loss to this envelope. Heating a cubicle of air to accomplish this task is a recent development. The earliest step in keeping the human body warm, which became necessary once our ancestors ventured out of the tropics, was the use of clothing. Clothes were the first passive heating system. A house can be thought of as an oversized union suit, enlarged to encompass the whole family.

Prior to the use of shelter, the next development after clothing for protecting the body was the discovery of fire. A campfire does a negligible amount of heating of the air, but its radiation will toast you nicely if you huddle closely enough, even when the temperature is below freezing. Fire is derived from solar energy, chemically stored by plants in the form of wood and fossil fuels.

The third step was to take fire inside a structure. The history of human shelter is the history of refining the combination of a shelter protecting against wind and rain with a fire inside for cooking and warmth. Shelters shield the body from convective heat loss: still air surrounding the body acts as an insulator. We all know how drastically wind speed affects the chill factor.

Caves were often used but caves are not always located where required. Other shelters were built of wood, plant materials, and stone. Some ancient cultures employed solar heating principles to supplement fire. At Mesa Verde the Pueblo Indians built cliff structures that capture the winter sun but are shaded in summer. The rocks retain heat accumulated during the day for use at night. These cliff dwellings were much more sophisticated structures in terms of solar energy use than are our suburban bread boxes.

Home heating has evolved chiefly in terms of increasing the convenience of fire rather than in using the sun's energy. The first fireplaces were smoky and inefficient, often cooling by draft more than heating. With the development of smelting came the iron stove that radiates heat from all sides. Stoves require a smaller draft and the draft can be controlled effectively, thus reducing cold air currents.

With the discovery of glass, which became available in quantity only in the late 1800s, some use began to be made of heat gain from the sun through windows. New Englanders were well aware of the importance of grouping their windows on the south. But since their houses lacked a means of storing the heat thus collected, fire remained the major heat source.

Increased affluence accompanying the Industrial Revolution led to larger houses with more rooms and doors for privacy. Doors prevented a single stove from heating the whole house, resulting in a stove or fireplace in every room. This led to the development of central heating. By putting a single stove in the basement and carrying hot air to each of the rooms through ducts, heat from a single fire could be distributed throughout a larger space. Thus we arrived at the "controlled environment" in which a coal, oil, or gas furnace governed by a thermostat warms our homes by an unseen hand so that we can pretend we are once more back in the tropics.

Now, at the end of the twentieth century, we are beginning to see a new approach to home heating, partly in reaction to the trend toward artificiality that has left us estranged from our natural environment, and partly due to the imminent depletion of fossil fuels.

Solar-Heating Systems

Compared with traditional heating systems, solar energy is technology on a human scale. An oil-fired furnace with its ductwork appendages often resembles a metal octopus. Compare this with the simplicity of sunlight heating a greenhouse. Of course, solar systems can burgeon into something just as intimidating as ordinary heaters, sometimes more so, but this happens only when they're intended to imitate the old systems in providing a completely controlled environment.

Three stages are involved in heating a solar home: collection, storage, and distribution. In conventional systems heat is produced by burning fossil fuels; in solar systems it is collected from the sun. Solar heat must therefore be stored for use at night and on cloudy days. It then must be distributed around the house to warm the rooms, just as in conventional systems heat is moved from a furnace through ducts to registers (forced-air systems) or from a boiler through pipes to baseboard radiators (hydronic systems).

As a matter of convenience, solar heating systems are divided into three types: active, passive, and hybrid. An **active** system is defined as one in which mechanical means are used to move heat from the collector to storage and from storage to the rooms; a **passive** system as one in which no supplementary mechanical energy is introduced; and a **hybrid** as one in which artificial means are used either to move the heat from collector to storage or from storage to rooms, but not both. In a hybrid system a small fan might assist distribution of heat, but passive systems rely solely on natural processes such as gravity and convection.

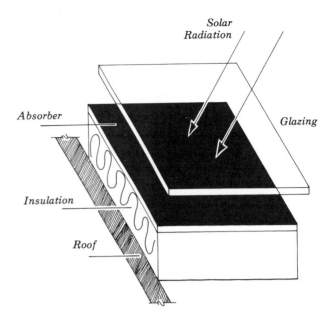

A flat-plate collector, simplified.

Active Systems

Most active systems employ a flat-plate collector. This consists of a box with a black bottom and a transparent top. Radiation from the sun passes through the cover, which consists of one or more layers of glazing (glass or plastic), and heats the *absorber*, a metal plate painted black to help absorb heat. The heat is transferred from the plate to either water or air.

In the simplest liquid collectors the metal consists of corrugated aluminum with water trickling down the valleys of the corrugations. The first house with a trickle collector was built by Harry E. Thomason of Washington, D.C., in 1959. More

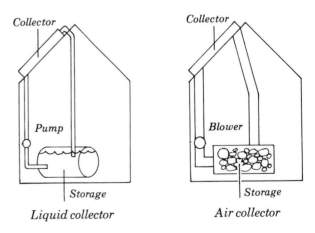

Liquid collector   Air collector

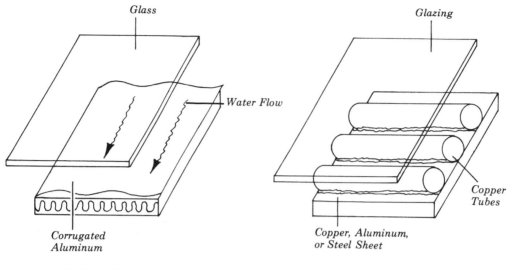

Glass

Water Flow

Corrugated
Aluminum

Trickle collector

Glazing

Copper
Tubes

Copper, Aluminum,
or Steel Sheet

Closed tube collector

sophisticated liquid collectors contain water within tubing bonded to a copper, aluminum, or steel absorber. In most systems the water automatically drains out to prevent freezing on cold nights; in some, antifreeze is added. A small pump circulates the water from the panels to an insulated storage tank and back to the collector. Air systems employ blowers to move the air through ducts to an insulated storage bin of rocks, which are warmed by the air. A thermostat capable of sensing when the collector is receiving enough heat to warm the storage triggers the pump or blower.

When the house needs heat, another thermostat activates the distribution stage. With an air system a second blower sends hot air through conventional ducts to room registers. A baseboard hydronic system may be used with a liquid collector, but typically the heat is instead transferred from water to air to permit use of forced-air distribution. With a trickle collector the storage tank heats a surrounding bin of rocks, and air blown through the rocks is ducted to the house. Other distribution methods involve a water-to-air heat exchanger, a coil of pipes resembling an auto radiator, located in the ductwork. Hot water is circulated through the pipes, and air is warmed by being blown over the coil.

Although each system must be designed for its particular site, broad generalizations can be made about the optimum size, positioning, and angle of collectors. One such rule of thumb would have the flat-plate panels equal one-third to one-half the floor area in order to obtain two-thirds of the space heating needs. Climate, house design, and collec-

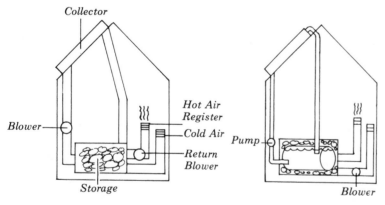

Collector

Blower

Hot Air
Register

Cold Air

Return
Blower

Storage

*Forced-air distribution
with air collector*

Pump

Blower

*Forced-air distribution
with heat exchanger*

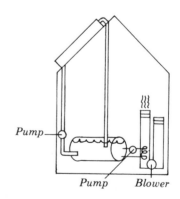

Pump

Pump   Blower

*Forced-air distribution
with trickle water collector*

tor efficiency all affect the accuracy of this generalization. Another rule calls for the collector to face due south, although up to 30° either east or west only slightly reduces performance. The panels should be mounted at an angle equal to latitude plus 15°, but once again some variation is acceptable.

The systems we've been describing, called **flat-plate collectors** because of their absorber sheets, perform best under direct sunlight, but also efficiently under conditions of diffuse radiation where radiant energy is available but diffused by cloud cover or air pollution. Just as you can be sunburned on a cloudy day, flat-plate collectors gain a surprising amount of heat when the sky is overcast.

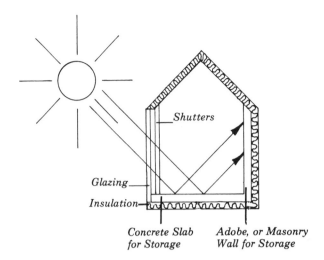

Direct gain.

Passive Systems

In passive systems solar energy is most commonly admitted by south-facing glazing, stored in walls and floors having sufficient thermal mass, and distributed by convective air flow. Materials such as concrete, brick, adobe, stone, or water stored in containers provide mass. A wood frame house with picture windows facing south gains solar heat during the day, but stores little for later use. We may include such houses in the passive category, but must recognize that storage in the form of thermal mass is necessary to make them solar homes.

Where the living spaces are directly heated by the sun, the system is called **direct gain.** With floors and walls providing storage, the heating system is inherent in the structure itself and adds little to construction costs. Other passive homes use some form of **indirect gain.** The sun's rays, rather than directly heating the rooms, heat an absorbing surface that in turn heats the house.

In this category are Trombe walls, invented in France in 1965 by Dr. Felix Trombe. A Trombe wall, built of concrete, stone, or adobe painted a dark color, and covered with glazing a few inches from the surface, reaches temperatures higher than would be acceptable in a direct-gain system. The wall prevents the occupants from directly experiencing the heat. Hot air on the surface rises and enters the room through a slot at the top of the wall while cooler room air drops and returns

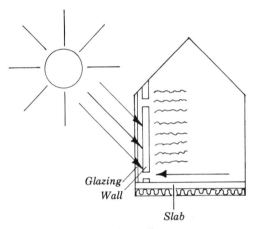

Indirect gain—Trombe wall.

through an opening near the bottom. Heat also reaches the house through the wall itself after a time lag of several hours, a fortunate arrangement since it is then available in the evening when needed.

A relative of the Trombe wall is a greenhouse with a massive north wall containing vents to siphon heat into the house. Often a fan will be added to improve air movement, and then the system will be classified as a hybrid. Other variations include walls composed of stacked water containers such as steel drums, bottles, or tubes; plastic bags of water exposed to the sun on flat rooftops; and water stored in bags or bottles in an attic that has a transparent south roof. Sometimes a flat-plate collector is used passively; that is, the heat is thermosiphoned into the house by natural convection rather than being transported by mechanical means.

5

The "Best" System

We'll be looking at examples of various kinds of active and passive collectors as well as hybrids and earth-integrated buildings. Visitors to our two homes invariably ask which system we prefer. Implicit in the question is an assumption that solar research will lead ultimately to a single, universally appropriate system. But just as people of today in different parts of the country select gas, oil, or electricity depending on cost and availability, solar homeowners will choose their systems depending on whether they're building a new house, combining solar heating with air conditioning, retrofitting an existing home, or building a multifamily dwelling. Available funds and individual preferences will figure in too.

Is there any point in investing in an expensive collector? Such collectors are economically viable, it is often argued, if all the hidden savings are added up for the lifetime of the system. "Life-cycle costing" does focus attention on valid considerations—the escalation of energy costs, the effect of inflation on reducing mortgage costs, and the reduction of utility bills allowing higher mortgage payments—but nevertheless for most of us a $10,000 solar collector is simply not cost-effective. Fortunately the "el cheapo" model often performs as well or nearly as well as the deluxe job. Even if you opt for a "high-tech" system, you'll find they vary enormously in cost: the Housing and Urban Development (HUD) 1976 solar demonstration program found prices of active systems running anywhere from $5,000 to $19,000. No one has proven that the high-priced systems are worth the extra money.

Quite acceptable alternatives exist, such as using passive techniques, shopping around for a reasonably priced site-fabricated collector, or building one yourself. In each case thorough research is your best safeguard. Some people advocate high-technology approaches on the grounds that low-tech systems sound too easy, tempting the do-it-yourselfer to dash into it. One professor cautioned against creating "solar junkyards" littered with broken-down collectors. But to advocate complex systems, which are out of range for most Americans, is to say in effect that solar is not yet

here. We would like to see alternative energy available at all levels of society, and today the only path to democratization is encouraging the use of simple, low-technology systems.

Perhaps mass production may someday reduce costs, but as a nation we can't afford to wait. We must get started on the conversion to solar energy. If we delay while the oil runs out, it may become impossible to accomplish the transition. How will we manufacture the aluminum and plastics for solar collectors? How will we refine and transport the copper and glass? No alternative exists but to go with the systems that people can afford today.

It's time the government started a mass education program in low-cost techniques. Lacking a federal initiative, the individual can still go the solar route by taking the time to get a decent grasp of the subject. The risks, though present, can be mitigated by education: you can learn from the mistakes and triumphs of those who have begun the exploration.

People and Places

We hope in this book to destroy a few myths about solar homeowners. You don't have to be rich to build a solar home. Nor do you have to belong to a counterculture group. Nor be an engineer. We found solar homeowners in every imaginable occupation: nurses, teachers, lawyers, salespersons, college professors, architects, folk singers, doctors, accountants, scientists, farmers, corporation presidents, ranchers—and engineers.

They ranged from young couples just starting out to elderly people building retirement homes. The houses varied from 600 to 4,000 square feet and cost anywhere from $4,000 to $120,000. Styles were ultramodern, traditional, or funky. Often the only difference between a solar home and the split-level next door was the panels on the roof.

We found solar dwellings everywhere from Maine to California in all kinds of climates. Average degree-days are given in the technical tables to provide an objective way of comparing weather conditions and heating needs for buildings in different parts of the country. They give no indication, however, of the amount of wind a building is subject to or the insolation it receives. *Insolation*, though it may look like a typographical error, means the solar radiation received by a surface. For each house we include the mean daily insolation (in Langleys) for December.

Reasons for using solar heating vary. The obvious motive is saving on fuel bills. Some people choose solar heat to avoid adding to air and thermal pollution caused by burning nonrenewable fuels. Their financial commitment sometimes may be far greater than they could hope to recoup in their lifetimes. For them, building a solar home is a social statement. A few do it for the novelty and prestige of being the first on the block to own a solar home. Architects and contractors build to get personal experience. Most interesting of all are the backyard inventors fascinated with the new technology.

Commonly mentioned is a desire for independence and freedom from the utility companies and the dictates of the Organization of Petroleum Exporting Countries (OPEC). The dream of many solar enthusiasts is a world where the individual has control over his own energy supply, with production of energy decentralized instead of in the grip of monopolies. To others, security is the most persuasive argument, providing insurance against future shortages and price jumps. No one, however, can be wholly safe: if our neighbor's heating bills soar because of an OPEC decision, we solar homeowners will find ourselves hit by the resulting across-the-board inflation. We'll all be in deep trouble from the economic disaster inevitable with a $45 billion yearly expenditure for oil imports.

A sociological study of owner motivations would help the spread of solar heating by determining factors that cause its acceptance and things that deter potential buyers, such as high front-end costs and esthetic complaints. Despite such barriers to growth the solar industry seems to be booming. Multinationals like Mobil Oil, Grumman Corp. and General Electric are getting into it. The New York Stock Exchange lists more than 100 corporations with solar research divisions. With the recent passage of the federal energy bill allowing tax credits for installing solar equipment, sales are expected to grow from the present $150 million a year to $1.5 billion a year by 1985. Potential effects on the economy, especially employment, are unimaginable. Compared with capital-intensive centralized power production, installation of solar collectors is an enormously labor-intensive industry, relying on human muscles more than large infusions of capital. Labor unions are beginning to recognize that a job explosion is ahead for solar machinists, construction workers, metal workers, plumbers, electricians, and factory workers.

The industry may develop into an economic revolution to rival the impact of the automobile. Ideally the changeover should be more than a technological revolution: it should be an opportunity to reexamine our lives. Just as the automobile brought pervasive transformations in the American life-style, we can envision solar energy doing the same—and we welcome it.

Most books and articles stress that living in a solar home is just like living in a conventional house. The idea is to avoid alienating people by giving the impression that they must relinquish comforts. It's true that you can have a solar home that requires no more of you than setting a thermostat, but it will be expensive. Even apart from the money, there is something almost horrifying about the kind of artificial wombs that our society has conditioned us to expect. As solar homeowners ourselves, we'd like to see the new technology involve a human commitment.

A couple of owners we interviewed describe themselves as sun worshippers. One likens living in a solar home to sailing a boat. There's the frustration of seeing the sun fade—like a sailor watching the wind drop. Waiting out a five-day cloudy spell can be like sitting out the doldrums. You can rail at it or eventually just get in tune with nature and accept what comes. As we once wrote about one of our homes, sun dwelling fosters "a closeness to the elements unusual in our mechanized civilization."

Why try to fit something so beautiful into old patterns? We'd be better off as a nation if we took the time to notice the weather and adjust our homes accordingly, closing window shutters and setting the thermostat back when a cold spell hits. Better off physically and somehow morally if we used this change in technology as a constant reminder of energy waste of all kinds and as a challenge to start reducing our level of consumption to our fair share.

Solar Homestead

The laws of ecosystem are not answerable to a criterion of success which necessarily includes the survival of man.

Stafford Beer

Why build a solar house? Everybody else is doing it? It makes economic sense? Being in the forefront of technological development is exciting? We see it as just one element in an attempt to live within the ecological realities of our planet. For the first time in history, we humans have developed the capacity to destroy not only ourselves, but the earth which is our home. If we are to find a less dismal prospect, we must conduct a reexamination of every aspect of our culture and personal lives. Building a solar house is not the solution; it is only the beginning of fundamental change—change in the very basis of our subsistence.

Not all solar homes are environmentally desirable. Some, expensive and employing scarce re-

sources, end up consuming more energy to build than they conserve in the long run. Steve, as a farmer, and I, an anthropologist and potter, wanted to construct a low-technology structure using recycled materials. Looking for integrity in form as well as materials, we turned to the kiva (ceremonial center) of the Pueblo Indians for the form of the solar addition. We started with a 1901 cabin that had plenty of character and seemingly timeless walls of juniper logs.

Steve came from Denver to western Colorado in 1972 with a group seeking to start a community, but within six months found himself sole proprietor of a 200-acre farm and a herd of goats. The old homestead cabin was in such disrepair that he considered tearing it down. Luckily some friends were not afraid to move in and do repair work, putting in a fireplace and rechinking the walls.

The Design

While figuring out how to turn the cabin into a solar home, Steve lived in a tipi and began accumulating materials by participating in the salvaging of an Air Force hangar outside Denver. One day in October a wind came up from the southwest, picked up the tipi and ripped it in half. The decision on the design of the addition had to be made. Having enjoyed the tipi shape, Steve wanted to repeat a round structure with a high pointed roof. Some medicine lodges were twelve-sided, and a dodecahedron with 8-foot sections would be easy to construct with conventional lumber. The northern four sections of the structure could be bermed with earth while the southern sections would contain solar collectors and windows. A dormer extending from the second floor loft would accommodate the master bedroom and storage space. Thus the matter of the design was settled; all that remained was

to build the house. Four years later, using recycled materials and doing it all ourselves, we are still finishing the interior while occupying most of the rooms.

I joined Steve just after he had gotten the roof on the addition. First he poured twelve concrete piers and stood a homegrown juniper post on each, connecting them at the top with pine lodge poles. The effect resembled a wooden Stonehenge, I'm told. Then a six-pointed star was built of 2×12's and raised on a temporary pole to support the rafters at the peak of the roof cone. One cold, windy day in October neighbors arrived to put up the rafters, fastening them with hickory pegs scavenged from old power poles. Through each of the rafters a ⅜-inch steel cable was threaded around the outside of the circle to create a tension band. The support was removed from beneath the star and the roof was sheeted with 2×6 tongue-and-groove fir from the hangar. The military gray paint was still on some of the roofing, and my first job was to stand on the top rung of an 18-foot orchard ladder to sand off the paint. I sincerely recommend refurbishing used lumber before you put it on a ceiling if you care about your neck.

TECHNICAL DATA

Owner-designer-builders: Steve Wolcott and Linda Lindsey, Colorado

General Features
Latitude: 39° N
Degree-days: 6,800
Insolation: 200
Heated area: 1,000 ft^2
Year of completion: 1977 (solar addition)
Insulation: Walls: 1" Styrofoam and 3½" fiberglass (kiva);
                          4"-9" urea-formaldehyde foam (cabin)
         Roof: 9" clinkers
Orientation: S
Solar system: Passive direct-gain and indirect gain

Collection System
Collectors: 84 ft^2 of double-glazed windows; 140 ft^2 of
   polyethylene glazing on greenhouse; 77 ft^2 thermosiphon-
   ing air collectors
Angle: 70°
Cover: Two layers of greenhouse glass
Absorber: Horizontally corrugated sheet metal, non-
   selective coating

Storage System
Container: 170 ft^3 wooden bin and 200 ft^3 crawl space
Material: Field stones, 4"-6" diameter
Location: Directly behind collectors and beneath floors

Distribution System
Natural convection

Auxiliary System
Backup: Two fireplaces
Fuel consumed: Three cords 1977–78

Domestic Hot Water
Insulated breadbox on roof with 30 gal. water in copper pipe
plus coil in fireplace

Costs
House: $8,000
Solar: $575
Domestic hot water: $400

Framing

On the north, with the exception of the sauna, the house was framed on the outside of the posts with 2×4's and plywood taken from the floor of the hangar—some with the linoleum still on it. Then came polystyrene, followed by tar paper. A 4-foot-high concrete wall was poured, and earth was pushed up to form a berm which covers the concrete and protects against the north wind.

Few windows were installed on the north. There is a beer bottle-cum-stained-glass window in the sauna, which extends out from the circle on the northernmost section of the first floor. The sauna's curved sides are double-thickness stone retaining walls filled with clinkers (the residue left from burning coal) for insulation.

Passive Flat Plate System

The four southern sections of the structure are dedicated to solar collection. Serving as the lower half of the wall, the collectors slope out at a 70° angle. Behind a double layer of old greenhouse glass in each collector is a piece of accordioned sheet metal painted black, with an airspace in

The old cabin still has a rustic look.

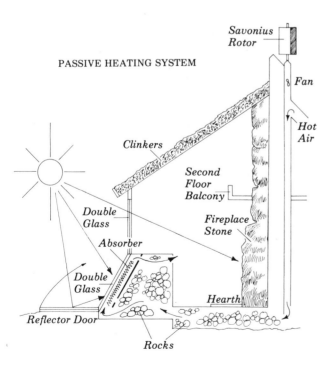

PASSIVE HEATING SYSTEM

Savonius Rotor

Fan

Hot Air

Clinkers

Second Floor Balcony

Double Glass

Fireplace Stone

Absorber

Double Glass

Hearth

Reflector Door

Rocks

front and back. Behind this absorber a 2-foot-wide wooden rock bin extends around the front of the living room. The shelf formed by the rock bin is used for growing plants, displaying pottery, and drying fruits and vegetables.

Polystyrene reflector doors with wood siding on the exterior, Masonite on the inside, and used printing plates as the reflective surface, fold up to keep the heat in during long winter nights, and protect the collectors during the summer days. They are manually operated from the exterior.

When the absorber is hotter than the rocks, air thermosiphons to them via gaps at the top of the wooden bin. Cool air returns to the collector through other gaps at the bottom. Manually operated registers allow warm air to flow into the house from the bin when needed. This arrangement allows us to have solar heat without complicated controls that are subject to breakdown. While it is quite simple, we've never seen another like it, although several houses have been built

The sun bounces off the reflector into the house.

time I am kept warm by the rays of the sun. On winter afternoons sunlight falls on the black-tiled hearth as well as the stone stairway and fireplace. Shaded by the shelf, couches along the south walls are protected from fading.

The finish work on the round addition was postponed last fall so that we could winterize the cabin, which was converted into a bedroom for our daughter, Kiva. We pulled out the rotten bottom logs and poured a concrete foundation. Because the juniper logs were requiring constant rechinking, we decided to cover the north and west walls with a rock face. First, though, we called in some friends who had just acquired a urea-formaldehyde foam outfit. It took them four hours standing in a cold drizzle to get the machine going, and the depth of the foam varies from 4 to 9 inches, but they gave us a special experimental rate. For a week the cabin looked like a gingerbread house with white icing, until Steve laid up basalt fieldstones to cover the insulation. With the foam on the outside, we haven't had a whiff of formaldehyde.

Clinkers on Roof

If you've been looking at the pictures of the house, you're probably asking, "But what is that stuff on the roof?" After putting a new roof on the cabin, Steve was thrashing around for a cheap means of insulating it without covering up the wooden ceiling. Finally someone suggested using clinkers. We don't burn coal, but clinkers are ubiquitous in the coal-mining valley below our mesa. Highly porous and not very heavy, they proved to be a good insulating material. On the cabin roof they held snow for up to three weeks, when inside temperatures reached 90° F. because of our oversized fireplace. We tossed clinkers on the kiva roof too, holding them on with redwood braces and adding interior braces to support the extra weight. In our arid climate we have little problem with ice buildup and hope the clinkers will extend the life of the asphalt roofing underneath by protecting it from the sun. We remain to our knowledge the only clinker insulators on the mesa. Anyway, if clinkers ever catch on, people will have to start paying for them instead of hauling them away from courthouses and schoolyards.

with thermosiphoning collectors below the house and storage under the floor. We have rocks under the floor too; hot air will be brought down to them from the peak of the ceiling by a windmill-operated fan on top of the cupola. If we find we need still more heat, we may install a fan to pull air from the top of the greenhouse to the rocks under the floors.

Putting the storage bin right in the living room does take up some 60 square feet of floor space, but we find the shelf extremely useful. Besides, the solid lower wall cuts off some incoming light so we experience no overheating or glare as we would if the front wall were entirely glass.

Direct Gain

Above the collectors are double-glazed windows offering direct solar gain in addition to the view. If I get up from my typewriter, I can look out at our Belgian workhorses grazing in the fields with Mount Lamborn in the background. At the same

The first winter in the solar addition, before construction of the greenhouse or reflector doors on the collectors.

Other examples of our recycling efforts include the use of aspen paneling from a Vietnam War–era bomb pallet factory, fir flooring bought at auction, and a coal stoker made into a fireplace.

Waterless Toilet

One item we bought new is our composting toilet. Even though it was manufactured in Sweden and came with a money-back performance guarantee, our county sanitarian was not enthusiastic about it, although it makes ultimate sense in an arid climate. During the drought of 1976–77, the mountain spring that supplies our water ran dry. Our Toa-Throne saves us 30 to 50 gallons of water a day, or about a quarter of our usage. We consider the heat loss from a turbine ventilator justifiable, since it eliminates odor. Although after a year we haven't yet had to empty the compost, the process does seem to be working; "flushing" with sawdust

as well as adding kitchen scraps helps to maintain the proper carbon-nitrogen ratio.

Bread Box Hot Water

The domestic solar hot water system is under construction now. There are basically two types of passive water heaters in current use: the thermosiphon type and the bread box. Because our roof configuration prevents us from placing a storage tank above the collector for thermosiphoning, we have a 2-inch copper pipe serpentined inside a long, low box that will be mounted on the north side of the roof in such a way as to catch the summer sun. The lid to this bread box is a reflector-insulator like the ones used on the space-heating collectors. The pipe, at once collector and storage, holds 30 gallons of water. It refills automatically from the mountain spring when hot water is used. In addition, a 30-gallon copper tank will store wa-

13

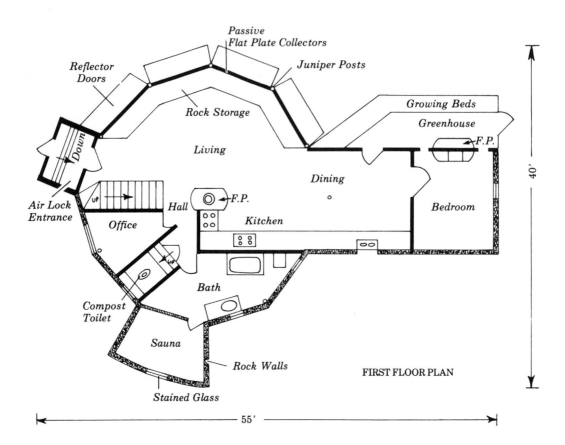

Passive
Flat Plate Collectors

Reflector
Doors

Juniper Posts

Rock Storage

Growing Beds

Greenhouse

F.P.

Living

Dining

Down

Bedroom

Air Lock
Entrance

Hall

F.P.

Office

Kitchen

Compost
Toilet

Bath

Sauna

Rock Walls

Stained Glass

FIRST FLOOR PLAN

40'

55'

ter heated by coils in the fireplace during the winter. Until these systems are hooked up, we are enjoying solar showers in the backyard from two 20-gallon drums flanked by reflector surfaces, and saunas in the winter.

The solar water heater is costing $400 due to the price of copper pipe—which cost $2 a foot. But we expect to pay for it through fuel savings within 2½ years. Our local supplier estimates that it costs the average family $144 a year for hot water using the propane-fired heater common in this rural area.

Keeping Costs Low

Our solar space-heating system, including the used glass at $1 a pane, came to $575. Just for fun I called a local firm and got estimates on a propane furnace—$1,000 installed—and an electric baseboard system—$500 on an unfinished house. The gentleman on the phone suggested that I try the electric system because "you can't tell what's going to happen with this energy situation, but it seems like we may have electricity around for a little longer than propane."

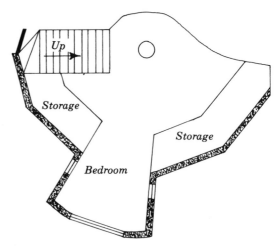

Up

Storage

Storage

Bedroom

SECOND FLOOR PLAN

We'll still have to burn some wood, but our fireplaces are designed for energy efficiency. Equipped with heat returns and outside air intakes, they have sheet-metal doors with thermostatically controlled dampers taken from junked Ashley stoves. The fireplace in the cabin is so effective that when we used to stoke it up to carry us through the night, we had to open the windows. Now we can easily heat the whole house with it. Last year we burned about 2 cords of wood in the cabin before opening up the wall to the solar-heated addition. After that we used another cord during the spring, for a total of about $60 in wood if we were to buy it rather than cut our own. During the same period, if we had been heating with propane, we would have used $400 worth of fuel because it was a mild winter; with increasing prices and more snow, the gas company estimates it would cost us $500 this season for space heating alone. Since the new central fireplace will be equipped for cooking, with an oven and swing-out grates, we expect to do most of our heating with it this winter, using a couple of cords.

Growing our own food is another important aspect of our attempt to live in tune with the environment. In addition to solar heat, the greenhouse attached to the cabin provides year-round vegetables. On 50 acres of irrigated fields we raise small grains—oats, wheat, millet, barley—and alfalfa, with the help of the horses and a tractor.

Looking out the south-facing windows in summer. The roof overhang keeps the sun out until it is needed in the fall. Beneath the shelf in the foreground is rock storage.

Our next project will be to remodel the root cellar and then build a bigger barn for the goats and sheep. After that we'll start on a machine shop-pottery studio, with apple cider bottles as a solar collection-storage wall. One thing about this environmentally sound living—it keeps you busy.

A Sort of a Saltbox

Experience is the name everyone gives to their mistakes.

Oscar Wilde

It's easy to tell you the mistakes made on this house because I worked on every inch of it from foundation to roof to finish work, learning in the process that I was not particularly fond of carpentry but also discovering the satisfaction of building my own home.

We did make mistakes such as using plastic pipe for a feeder at the top of our collector. Before the system could be put into operation, the intense heat melted the pipe into a piece of spaghetti. After replacing it with copper, we had more troubles. A windstorm ripped off a large piece of the collector's fiberglass skin before it was permanently attached. Then the rocks for the heat storage were mislaid. A local quarry was to deliver them in time for a rock schlepping party: two dozen volunteers

had agreed to form a human chain to hand the rocks from dump site to storage bin. Not the most elegant solution but workable. The rocks, however, failed to appear on the appointed day. Upon investigation we found that the truck driver, new to the area, had delivered them "somewhere" along our road, but no one at the quarry knew exactly where since the driver had immediately left town. The schlepping was cancelled, and an entire day was spent driving up and down our road and all the roads in the township with a representative from the company. The search was unsuccessful, so presumably someone somewhere still has 20 tons of unwanted rock piled among his pine trees.

Most of all we had problems with financing.

Money was hard to come by, particularly for a solar-heated home. In 1974 fewer than four dozen existed in this country, and none of those was in our state of Maine.

Down east banks are notoriously conservative, and our local establishment leads the pack. Two days after submitting our design we were informed the bank directors had just voted to discontinue all construction loans until further notice.

We then tried nearly every bank in Maine, and a couple in Massachusetts, without success. Finally on a hunch we called the editor of a paper sympathetic to environmental causes, and he put us onto some people in the banking world who he felt would be friendly to an alternative energy project.

They were friendly enough after we had endured a third degree and agreed to subsidize inspection visits by an architectural firm to the tune of $600. In the end we received the first conventional mortgage granted by a Maine bank for solar housing, probably one of the first in the nation.

Admittedly, part of our financing problem was due to our amateur status. Our designer, Ernest McMullen, a longtime alternative energy enthusiast, is a self-trained architect who at the time had designed only one other home. My philosopher husband, Dick, had left the security of a tenured position at a southern university, joining Ernie on the faculty of an experimental college devoted to the study of ecology. Also trained in philosophy, I had no more building experience than Dick.

Active and Passive Systems

In retrospect it's surprising we made as few mistakes as we did. We lucked into an efficient combination of solar heating and energy-saving features: double glazing and heavy insulation, a low

---

TECHNICAL DATA

Owner-builders: Richard and Norah Davis, Maine
Designer: Ernest McMullen

General Features

Latitude: 44° 30' N
Degree-days: 7,900
Insolation: 140
Heated area: 1,500 ft^2
Year of completion: 1974
Insulation: Walls: 7" fiberglass
          Roof: 9½" fiberglass
          Floor/foundation: 3½" fiberglass and 2"
                    Styrofoam
          Shutters: 2" Styrofoam
Orientation: 10° W of S
Solar system: Hybrid—active liquid drain-down and passive direct-gain

Collection System

Passive collector: 190 ft^2 double-glazed windows
Active collector: 520 ft^2 trickle collector
  Angle: 54°
  Cover: Double layer Kalwall Sun-Lite®—outer 0.040", inner 0.025"
  Absorber: Corrugated aluminum, nonselective coating, 2" fiberglass
  Pump: 1/3 hp

Storage System

Container: Steel tank enclosed in 16'×8'×7' wooden bin lined with plastic

Material: 1,600 gal. water and 20 tons rock up to 9" diameter
Location: SW corner of basement
Insulation: 6" fiberglass

Distribution System

Natural convection or 1/3-hp blower controlled by manual switch introduces air into bottom of rock bin where it is heated by contact with stones as it rises to an exhaust port leading through insulated ducts to room registers.

Auxiliary System

Backup: Fireplace and 85,000-Btu wood furnace (heat stored in rock bin) and one electric wall heater
Fuel consumed: Three cords at $55/cord and 400 kwh at 4 ¢/
Angle: 54°

Domestic Hot Water

Thirty-gal. tank immersed in storage tank preheats for twenty-gal. electric water heater.

Costs

House: $36,000
Solar: $1,812
Operating costs: Less than $20 a year
Maintenance costs: $10 to replace a sensor

---

northern profile with only one window, a southern facade with room for a trickle collector as well as a greenhouse and sliding glass doors for passively gathering solar heat, and an open floor plan facilitating dispersal of that passive heat.

Built on a rock ledge that supported few life forms other than lichen, the house had a minimal impact on the environment. The greenhouse was integrated into the design rather than being just stuck on as some greenhouses seem to be. Inside, a free-standing ceramic fireplace was equipped with two dampers and a sheet-metal cover to reduce heat loss. We used as many recycled materials as we could find, from a used steel tank for heat storage to quarter-sawn oak flooring, solid pine doors, and cypress counter tops. We scraped the rust from the steel tank, discarded from a gas station, and added two coats of fiberglass. For access, a manhole was cut in the top to correspond to a trap door in the floor above.

Lacking instrumentation, we have only rough measurements of performance. One winter we got a reading of 117° F. by holding a thermometer against the feeder pipe where it leaves the tank. A duct leads from the wood furnace into the rock bin, but since the furnace had stood idle for several days prior to our measurements, it did not affect the reading. When we use the furnace, we burn it hot to avoid creosote buildup, and the excess heat is retained in the rocks for a day or two.

In 1977 – 78 our total heating bill was $180, one-fourth the oil bills of coastal Maine houses of similar size. Insulation accounts for a good part of that savings, but the sun—through both active and passive gain—meets 70.7 percent of our actual heat load, according to a study done by graduate students at Worcester Polytechnic Institute. The true percentage may be a little lower as some figures in the thesis look questionable to us—yet what matters is our $600 yearly savings.

Because of a tight budget, a very tight schedule, and a lack of readily available information, we made errors that are avoidable today. The house is built in an area devastated by a 1947 forest fire, and the secondary growth is too small to block the northern gales. We have a good southern exposure but excessive infiltration. Our main entrance is on the north side with no vestibule to serve as an air

Dining area, kitchen, and greenhouse. The ceiling and cabinets are ordinary No. 3 grade tongue-and-groove pine. The floors are tongue-and-groove oak, quarter sawn, a real find. They are from the ballroom floor of the Evelyn Walsh McLean mansion. She was famous for owning the Hope Diamond, now in the Smithsonian. (All photos of this house courtesy Claude Bolduc.)

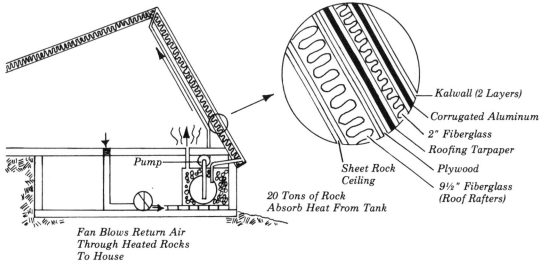

Kalwall (2 Layers)
Corrugated Aluminum
2" Fiberglass
Roofing Tarpaper
Plywood
9½" Fiberglass
(Roof Rafters)

Sheet Rock
Ceiling

Pump

20 Tons of Rock
Absorb Heat From Tank

Fan Blows Return Air
Through Heated Rocks
To House

Schematic of Davis heating system.

lock. The chimney is relegated to the outside of an exterior wall, rather than inside in a central location where we could gain heat from it. As often happens, we built larger than we really need: I am now convinced the two of us and our cats would be happy living in 1,000 square feet. Worst of all, one room sticks out baldly from the rest of the house with three of its walls exposed to the elements. Almost as bad, our high ceilings mean that we're heating unusable volume. If I had it to do over again, I'd make the whole building more compact and lower that pretty ceiling in favor of an attic stuffed with insulation.

Pump Problem

Had our budget been looser, we might have made the collector a little fancier. An aluminum sandwich with water trickling between the two layers would be more efficient. A bronze or stainless steel pump would be better. Right after installing one made of cast iron, we learned that in open systems oxygen mixes with the water providing ideal conditions for rust. We might have used glass covers, since our Sun-Lite® (polyester-filled fiberglass) sometimes rumbles in high winds. The noise problem could be remedied by adding trim, but more trim would mean more shading on the collector—reducing the net collecting area. Because we had

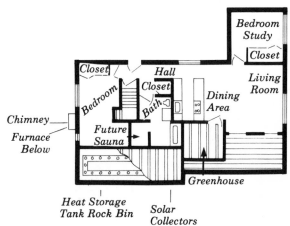

Bedroom
Study
Closet

Closet
Hall
Closet
Living
Room

Bedroom
Bath
Dining
Area

Chimney
Furnace
Below

Future
Sauna

Greenhouse

Heat Storage
Tank Rock Bin
Solar
Collectors

FIRST FLOOR PLAN

trouble getting a tight seal between the plastic and trim, steam escapes at the top and we lose more heat. We lose additional water at the bottom from condensation dripping down, but recently we modified the trough at the bottom to stop overflow. All this water loss was adding up. Two or three times a year we had to refill the storage tank by hooking a hose to a faucet, although this should be greatly reduced now that we've repaired the gutter to eliminate drips.

Not long ago Brookhaven National Laboratory of Long Island, New York, sent a technician to examine the Sun-Lite on our collector and greenhouse to determine how well it has stood up to 4½ years of Maine weather. He found that two layers of Sun-Lite, when new, are virtually identical to Thermopane™ in the number of Btu transmitted. After exposure to sun, wind, and rain, the Sun-Lite covers on the greenhouse have degraded 14.3 percent. The instrument used for the test has an error factor of plus or minus 3 percent. Assuming a similar deterioration on the glazing of the collector and assuming no significant worsening, Sun-Lite seems a viable alternative to glass, considering cost and ease of installation.

We recently had our first maintenance expense when our differential thermostat went berserk. Among other behavioral peculiarities, the "brain" of our system took to turning on the collector pump at night. A differential thermostat has two sensors, one on the surface of the collector and the other in the storage. Comparing the readings of the sensors, the thermostat triggers the pump when the absorber becomes warm enough to heat the tank, and in the afternoon when the surface cools down, turns it off. Our troubles were found to be caused by a sensor gone bad, a $10 repair job.

A few dollars now and then for maintenance is painless when your heat bill is gentle on the budget. We expect to do some tinkering, a hobby common enough with solar do-it-yourselfers. When we

Shutters in the living room fold like an accordion; those in the greenhouse open and close in pairs. We found that they had to be fully weatherstripped and tightly latched to be fully effective. Getting a tight seal all the way around is important.

DAVIS HOUSE: COSTS FOR SOLAR AND AUXILIARY HEATING SYSTEMS

Solar heat system:

Plans for Thomason collector	$ 30.00
Sun-Lite glazing material	355.00
Aluminum roofing	180.00
Storage tank and domestic water tank (both recycled)	110.00
Black paint	126.00
Pump	100.00
Concrete block	56.00
Lumber and plumbing (some recycled)	160.00
Stone	75.00
Materials	1,192.00
Labor (semi-skilled 200 hours)	600.00
Total	1,792.00
Replacing plastic pipe with copper	20.00
Total	$1,812.00

Auxiliary system:

Wood furnace and plenum	$ 580.00
Blower (recycled)	10.00
Ductwork (salvaged)	10.00
Ceramic fireplace	450.00
Total	$1,050.00

get around to replacing that pump, which besides being iron is larger than needed, we'll move it to the basement. Right now it lives in a future sauna next to the bedroom where its friendly hum lets me know when the day is sunny without my getting out of bed or even opening my eyes. That's fine most mornings—except when I'd like to sleep late.

It is nice knowing what the day is going to be like. Being a sun dweller means caring about the weather. I'm worse than the worst weather complainer: I grumble at long cloudy spells, at the brevity of December days in this far northern clime, at the passing cloud that causes a perceptible drop in room temperature. But I have found a new appreciation for the lack of smog in Maine, and a snowfall now means more to me than a chance to go cross-country skiing: snow, a good reflector, intensifies the sun striking the collector.

Conscious of Waste

Living in a solar home is making me increasingly conscious of energy waste to the point of fanaticism. For one thing I hurry guests in the door to avoid heat loss. For another, each year I lower the

Living room fireplace is the backup that we use almost all the time in preference to a wood furnace in the basement. Although nowhere near as efficient as a good wood stove, the fireplace does have a sheet metal cover and two dampers to improve efficiency. The two-inch thick silicon carbide walls help retain heat. If we had an outside air intake, the fireplace would be much more efficient. Alternatively, an airtight wood stove would get us through the winter with one or two cords, instead of the three we now use.

thermostat another degree or two. My comfort threshold used to be 70°, but now it is 65° during the day and 55° for sleeping. I rush to close our shutters when a sunny day turns to clouds. It takes 135 seconds, but the installation of shutters cut our heating bill nearly in half. Using our new carpentry skills, we constructed them out of Styrofoam sandwiched between Masonite covered with burlap. Since they partially block the windows, heat gains were slightly reduced but at the same time excess heat buildup can now be controlled. Shutters have added a twice-daily renewal of my awareness of the weather.

Creature Comfort

It would be unfair not to mention the bad side of living in a solár home. For us, burning wood as a backup used to be an inconvenience, but as we added improvements like shutters, the furnace in the basement has become less and less of a nuisance. Now it's stoked only fifteen or twenty times a winter. Despite the inherent inefficiency of our living room fireplace, we find it's all that is needed

to take the chill off except when the wind is heavy or the temperature falls below zero. I'm embarrassed to admit that we finally decided to put in an electric heater for our cats to huddle around when we're traveling away from home, but it adds less than $20 to our yearly electric bill.

Sad to say, the hardest part of living in a solar home has been the stream of visitors. Having had an estimated 2,000 solar enthusiasts knock at our door, we've become a bit weary at times of acting as tour guides. On the other hand, there have been visitors who have lightened our day, particularly the poultry farmer who wanted to build a solar-heated chicken house.

Some have been rather bizarre. One night shortly after we moved in, a corpse in a box was dumped in our front yard. The next day we watched with natural apprehension as police swarmed through our woods looking for clues, and we waited through an afternoon for them to get around to questioning us. When detectives did finally come to our door, they asked two or three disappointingly routine questions. But make no mistake, they wanted all the facts on the solar collector.

21

Solar One

*The United States and much of the free world have been living
high off accumulated capital energy in the cream of the coal, oil,
and gas reserves.*

Palmer Cosslett Putnam

To the best of their knowledge, Bob and Statia Ramage's home was the first privately built solar house in New York. Located in the Hudson Valley between the Catskills and the Berkshires, the Ramage residence is well situated for solar heating. With the lower floor partly built into a south-easterly slope, the house has both southern exposure and a hill for protection from north winds.

Customarily only the main floor is heated. The basement contains a workshop and garage, the solar storage, and an office heated when in use with a museum-piece wood stove that we immediately coveted. The main level contains bedrooms,

the living area, and a glassed-in porch. Above the bedroom half of the house is an attic dormitory, which the Ramages heat only when the grandchildren are visiting. Bob Ramage is a retired commercial artist and Statia a retired social worker.

Pioneering Design

Bob refers to their solar consultant, Neil Freer of Ahead, Inc., as a "dedicated engineer" and a "good friend," but says that neither he nor Neil can claim to have made no mistakes. Although he's unhappy with the end results in some ways, Bob notes that they were doing their planning in 1974 when there were few answers available. "In fact there are still very few definitive answers," he says. "People are still experimenting."

Neil built the air collector out of recycled offset printing press plates, forming them in his shop and then assembling them at the site. Each panel fits between two roof rafters. Bob refuses to hazard a guess as to the total cost: "We did the collectors ourselves, so there's no price per square foot that anybody can pin down unless they really go to work on all the labor, the delays on account of rainy days, all that sort of thing."

Hybrid System

Like many of us with active collectors, the Ramages like to think of their home as having a hybrid system since much of their heat during daylight hours comes from passive gain: half the windows are on the south side of the house and on sunny days the porch never needs supplementary heat. Both Bob and Statia believe that passive is the future of solar heating.

Although their system suits their needs, Bob characterizes it as "almost a blind alley" because of its complexity. It has no fewer than seven (or eight depending on how you count) operating modes. In Mode A air is blown through the storage bin to the collectors and back in a closed loop. Heat is being stored with no call on it from the house. In Mode B the thermostat signals for heat, so a blower sends solar-heated air to the rooms. When not enough solar heat is stored, the electric auxiliary in the main duct kicks on (Mode Bx) and warms the air to the desired temperature.

Modes C and D are cooling modes. In Mode C the blowers bring cool night air from a vent beneath the porch into the storage, and in Mode D the cooler air is circulated the next day as needed. The Ramages have not operated their cooling system for they're high enough on their hill to receive

---

TECHNICAL DATA

Owner-designers: Statia and Robert Ramage, New York
Solar consultant: Neil Freer

General Features
Latitude: 41°30′ N
Degree-days: 6,000
Insolation: 115
Heated area: 1,550 ft^2
Year of completion: 1976
Insulation: Walls: 4" urethane and urea-formaldehyde foam
          Roof: 7" fiberglass
          Foundation: 6" fiberglass
          Shutters: Insulated drapes
Orientation: 12° W of S
Solar system: Hybrid—active air and passive direct-gain

Collection System
Active collector: 1,000 ft^2 gross, 722 ft^2 net
  Manufacturer: Ahead, Inc., West Hurley, NY
  Angle: 60°
  Cover: Double layer Kalwall Sun-Lite®—outer 0.040", inner 0.025"
  Absorber: Aluminum sheet, nonselective coating, 3½" fiberglass

Storage System
Container: 17½' × 12' × 8' poured concrete bin
Material: 1,640 gal. water in glass bottles

Location: Basement
Insulation: 3" urea-formaldehyde foam

Distribution System
Two ½-hp blowers send air upward behind the absorber at 4,000 cfm. Hot air collected in upper heater flows downward in a duct beside the central chimney. Room air circulated among bottles by one of the two blowers.

Auxiliary System
Backup: 13,000-watt electric resistance and two wood stoves
Fuel consumed: 3,641 kwh at 5 ¢/kwh average and 2-2½ cords at $20-25/cord

Domestic Hot Water System
Collector: 96 ft^2 with antifreeze
  Angle: Three panels, one stationary at 90°, the other two movable from 90° to 30°
  Cover: None (located inside greenhouse)
  Absorber: Factory-made aluminum absorber with integral channels
  Pump: ¼-hp
  Storage: Heat exchanger inside 80-gal. tank, boosted by 80-gal. electric water heater

Costs
House: $96,000
Maintenance costs: $5 to replace filters

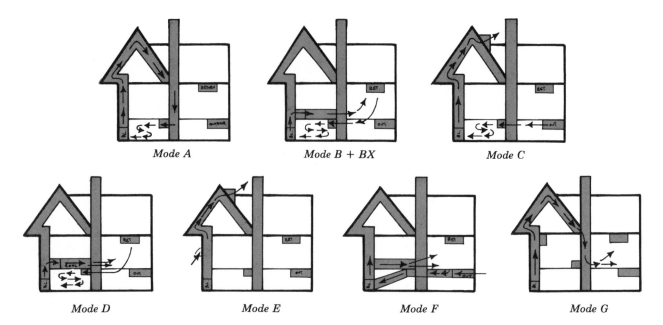

Mode A   Mode B + BX   Mode C

Mode D   Mode E   Mode F   Mode G

The seven operating modes of the Ramage system.

afternoon breezes. With the glass doors open the porch becomes a delightful screened-in sitting area. Like many solar dwellers Statia finds ordinary air conditioning unpleasant and artificial.

Mode E is for dumping excess heat from the collector. Air is drawn by natural convection through a vent at the bottom and out an opening at the roof peak. A sensor opens the vents when the collector hits 172° F. Although their function is to protect against summer overheating, they have been known to open on a bright sunny day in February when the house had no need of heat and some gauge somewhere was a little off, preventing the collector from being cooled by heat drawn off into storage. With nowhere else for the hot air to go, Mode E took over and vented it—so that Bob was dismayed to discover his collector trying to heat the whole county.

Mode F is nothing more than a fan. Bringing fresh air directly into the house, bypassing storage, it's like an attic fan but located in the cellar instead. Then there is Mode G, an afterthought added for sending solar-heated air straight to the house, again bypassing storage.

Complicated Controls

Needless to say, the controls for all this are complicated, what with sensors and relays turning on fans and starting tiny motors that open and close each damper depending on the mode. In a conventional house the furnace operates in only one mode and delivers air at only one temperature. When the furnace is on the roof, however, there has to be a supplementary system, which means different rates of airflow. As we'll see in the Sanford house (see p. 42), if you use fewer modes an air system need not be this complicated.

Computer and other monitoring equipment in basement.

The Ramage residence also has zone heating, a good idea in that, like task lighting, heat is provided only when a given area is in use. The guest room is one zone, the living room another, and so on. "Theoretically it's more efficient," Bob says. "I guess it works but when you're dealing with forced-air, the dampers have to be almost perfect." If they leak here and there and you have a dozen of them all doing a little leaking, it all adds up to a loss of pressure. Bob is considering reducing the number of zones from six to three, thus eliminating a few sensors and dampers.

Rocks First ...

Statia has some heartfelt advice to pass along: "We want everyone to know that you can't put just any kind of rocks in your storage." The Ramages started out with a rock bin, but thought they were losing heat to the ground through the bin's uninsulated floor. Since no one could think of a way to put in insulation without removing the rocks, Bob hauled out all 65 tons, first using an electric hammer to break through the foot-thick cement wall of the storage area. As he carried them out he discovered that much of his 1½-inch gravel had turned to dust, since it contained too much shale. The fine powder had been reducing airflow. What do you do with 65 tons of rocks? Bob bought a tractor with a loader and has been making driveways and paths around the house ever since.

... Then Bottles

Deciding to switch to water-filled bottles as a more efficient heat store, Bob built shelves out of wire screen—turkey wire he calls it—and put in baffling to form compartments to force the air to come into contact with every one of 1,640 gallon jugs. In all it travels nearly 50 feet to go 10 linear feet. Here's another piece of advice: don't buy plastic bottles for solar storage, at least not the kind used for cider. Bob's leaked at the weakest points as they bulged and contracted from temperature changes. Probably he overfilled them, but he later learned that bottles intended to store liquids for short periods eventually leak through the "pores"

Plastic jugs painted black and storing 150 gallons of water provide heat in greenhouse. Temperatures in the greenhouse go down to the low 30s, but the water in the bottles never freezes.

of the plastic. With that type of container, you could expect a 2 percent water loss per year even without seam leakage. Heavier plastic such as that used for Clorox bottles might work but would cost considerably more.

Bob emptied all 1,640 plastic bottles, which he had bought from a wholesaler, and went to The Bronx where he located secondhand glass bottles formerly used in drugstores for storing syrup. Some had been washed; others had a few drops left so that when filled with water interesting colors appeared—cola, chocolate, and even blue. Kaleidoscopic solar heat storage.

System Monitored

The area electric company has been monitoring the Ramage system for a year. Meters record the number of kilowatt hours used for resistance heating, operating the blowers, and boosting hot water. A computer is fed data on insolation, air velocity in the ducts, and house, collector, storage, and exterior temperatures. Visiting on a sunny day in early April, we watched through the computer's glass door as numbers flashed on the dials. Outside temperature: 52° F. Air temperature at the top of the collector: 146° F. Once, a three-year-old grandson got the glass door open, punched every button, and threw the whole thing out of kilter. The electric company had to make a service call to readjust it and install a door hook out of the reach of curious fingers.

RAMAGE HOUSE: ELECTICITY
USAGE, DEC. 12, 1976 TO DEC. 12, 1977

Total kwh	20,950	
Hot water booster	4,720	22.5%
Auxiliary heat	3,641	17.4%
Blower motors	1,323	6.3%
All other	11,266	53.7%

All electric, the Ramage home has a freezer, frostfree refrigerator, dishwasher, washer and dryer, and computer. Rates per kilowatt hour are on a sliding scale from five cents down to a penny depending on how much the customer uses. This makes it hard to separate out the exact cost of heat, but we estimate that operation of the collector and resistance heaters runs approximately $200 a year. The house temperature is kept at 68° F., and the main wood stove is on most of the time. Although Bob and Statia cut their own wood, they figure that a winter's supply of two cords costs a total of $40 to $50 when you count buying and maintaining chain saws and hiring neighborhood kids to help. Before building this retirement home, the Ramages lived at the bottom of their hill in a big, drafty farmhouse that used $800 worth of oil a year. Today's $250 looks good to them.

Water or Air?

An architect friend of the Ramages once told them that if people choose a water system, they wish they had air, but if they choose air, they wish they had water. The Ramages opted for air to avoid running the risk of leaks and because the notion of using antifreeze was unappealing. You could install a drain-down system instead, but still there would be the fact that an air system can be an integral part of the roof, thereby reducing material costs. Utterly fair, the Ramages point out that

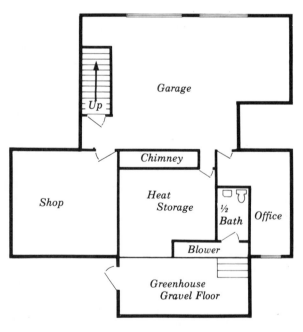

BASEMENT FLOOR PLAN

MAIN FLOOR PLAN

Living room with Jotul wood stove.

air requires more duct work, which is expensive. Even so, they believe that the long-term trend will be toward air systems.

Or passive. When we arrived the Ramages were engrossed in weeding a huge garden. As ardent gardeners, they've built a greenhouse on the south side of the house and are considering venting it into Bob's workshop for some passive gain.

Greenhouse Collector

On the back wall of the greenhouse is a collector for heating hot water. The company that manufactured it went out of business right after the Ramages bought a system, so "we were on our own from there on." Since the collector is inside a glazed structure, no cover is needed. Two of the three panels are hinged so they can be swung down for maximum heat gain from the winter sun, which is low in the sky, and up for summer sun. One panel is fixed because it would get in the way if it tilted.

Well water comes up at 55° or 60°, and the solar collector heats it to 95°. For the sake of the dishwasher, which requires hotter water, the electric heater is kept at 130° F. Bob considered dispensing with a standard water heater in favor of a small tank under the kitchen sink to boost water for the dishwasher, relying on solar-heated water for

The interior of the greenhouse, with the 150 plastic bottles, and two of the three domestic hot water collectors. Hinges permit two of the panels to be raised and lowered to follow the sun for maximum collection efficiency year round. The panels are vertical in winter when the sun is low on the horizon, tilted up for summer collection when the sun is high.

bathing and washing clothes where lower temperatures are acceptable. "But I got chicken and didn't have nerve enough to do it. I sort of wish I had."

Even so, the Ramages could justifiably be accused of being solar addicts—like most of us in sun dwellings who've found that one thing leads to another. For one, they put in a sliding glass door for the main entrance with a regular front door on the other side of an unheated foyer, creating an ingenious and attractive air lock. "That was one of those middle-of-the-night ideas," Bob says. For another, he took 150 of the plastic bottles, painted them black, and hung them in the greenhouse where leaks won't damage anything. At night heat stored in the water helps prevent the greenhouse from freezing. But then Bob may have outdone us all in solar one-upmanship: he painted the sink in the greenhouse black. Where will it all end?

Rip Van Winkle's Retrofit

By nature men are pretty much alike; it is learning and practice that set them apart.

Confucius

Howard E. ("Rip") Van Winkle and his wife, Ellie, woke up to the energy crisis in 1974. Being among the earliest solar retrofitters in the Boulder area, they feel an obligation to share their experience with others. On a blistering day in July we found Rip in his Bermuda shorts and straw hat, speaking through a homemade sound system to a tour group assembled on his front lawn.

"If you're considering doing a solar retrofit on your home," Rip says, "the first thing to do is— stop. Before you take any action on a solar system, think about all the possible ways to reduce heat loss in your structure." Insulate thoroughly, especially the roof and north walls. Have a furnace

contractor adjust your furnace for maximum efficiency—or do it yourself if you have reason to believe you know what you're doing.

Then after you've done everything you can to tighten up your house, set out to get the maximum amount of passive gain. A passive system, Rip argues, is cheaper and easier to live with than an active one. Because it serves multiple uses, adding a greenhouse is one of the best passive techniques. Alternatives include punching new windows into your south wall or glazing a concrete wall to create a Trombe collector.

Any home can get 30 to 40 percent of its heating from the sun with only a tolerable 10 percent overheating, says Rip. How much passive gain you can include is limited by how much thermal mass for storage you can add, because the energy not stored will either be wasted or will cause overheating. Storage does not necessarily mean rock beds or water drums; the walls and floors of an ordinary house will hold quite a bit of heat.

Step three in Rip's retrofitting program is the addition of an active collector to take care of domestic hot water. If you happen to have room for more collector than required for the water, you can use the excess heat to warm the house as the Van Winkles have done.

Insulate First

All the books tell you to insulate first, but solar enthusiasts are often too anxious to get down to the hardware. The Van Winkles' experience shows that cutting heat losses really makes a difference.

They bought their ranch-style home in the foothills of the Rockies twelve years ago and began retrofitting it in 1974. To the 3 inches of fiberglass insulation in the attic, another 8 was added. Even though the energy crisis was impending, the insulation contractor tried to talk Rip out of using that much. Blowing insulation into the walls was considered, but it seemed that adding one more inch of cellulose to the rock wool already there was not worth the cost. In the cold northwest bedroom Rip went to the trouble of tearing out the dry wall and putting in fiberglass. The basement walls were insulated too, as well as the chimney in the attic.

Next the windows were triple-glazed. Ellie had made drapes, but they were rarely pulled because the heat registers for the old forced-air system were located just below the windows. Closing the drapes directed the warm air toward the windows rather than into the house. A first set of storm windows was purchased in 1967 for $200. Seven

---

TECHNICAL DATA

Owner-designer-builders: H. E. "Rip" and Ellie Van Winkle, Colorado

General Features
Latitude: 40° N
Degree-days: 5,500
Insolation: 185
Heated area: 3,950 ft^2
Year of completion: House, 1966; greenhouse, 1976; active system, 1977
Insulation: Walls: 2" rock wool
         Roof: 11" fiberglass
         Basement walls: 1" Styrofoam
Orientation: 26° E of S
Solar system: Hybrid—greenhouse and active liquid drain-down

Collection System
Passive collector: 140 ft^2 greenhouse
   Angle: 20°, 85°
   Cover: Single glazed with double-strength window glass
Active collector: 150 ft^2
   Angle: 50°
   Absorber: Copper tubing with copper backing
   Cover: Single glazed with double-strength window glass
   Pump: 1/20 hp

Storage System
Container: Concrete cistern divided into three compartments

Material: 2,200 gal. water
Location: Under greenhouse
Insulation: 2" Styrofoam

Distribution System
Living area upstairs is heated by natural convection from greenhouse through living room windows. A pump forces water through a heat exchanger in storage tank to heating tank in basement. Greenhouse is heated at night by heat exchanger under planting bench and air blown through radiator.

Auxiliary System
Backup: Gas forced-air furnace
Fuel consumed: $92 in natural gas 1977-78 season

Domestic Hot Water
Water is preheated in heat exchanger in storage tank, then boosted by 40-gal. gas water heater.

Costs
House: $30,000
Insulation: $400
Triple glazing: $800
Solar greenhouse: $2,000
Active system: $1,200

---

years later Rip made frames for a third layer and had the glass put in for the same price.

Over the years other fuel-saving measures have included turning down the gas water heater and putting an insulating jacket on it. The Van Winkles also allow dust to collect on their automatic dryer as they've returned to hanging clothes on a line. Most importantly, they turned the thermostat down to 64° F.

The Greenhouse

Rip directs the attention of the crowd to a chart showing the result of these efforts. In 1966 heating the house required 225,000 cubic feet of natural gas; by 1975 gas consumption had plummeted to 100,000 cubic feet. Construction of the greenhouse brought it down to 75,000, while addition of the active collector further reduced it to 50,000 cubic feet annually. (Even more could be saved through a little tightening up of the solar system.)

Members of the tour group admire the greenhouse tucked into a U formed by the garage, living room, and south bedroom. Seven feet wide by 20 feet long, the plant room is framed with redwood. As an afterthought the strip of glass nearest the house was insulated with fiberglass to prevent

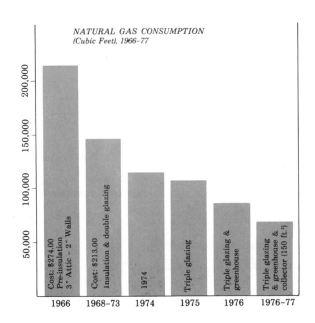

This fuel consumption chart shows his fuel bills have dropped from $275 a year to $100 due to insulation and solar heating.

heat loss and provide shade. Rip now regrets that they failed to double glaze the greenhouse. His attempt to cut heat losses by wrapping it in plastic was unsuccessful because the sharp corners tore the plastic. Adding a second "storm greenhouse" in front is a possibility he is considering, but not too seriously.

At the moment, however, the room is warm enough. None of the tourists stays inside for longer than a few minutes, for it is over 100° F. in there. The highest temperature recorded was 115°, Ellie says, when it wasn't much cooler than that outside. For air circulation, Rip designed high and low vents, but last summer a plague of grasshoppers descended upon eastern Colorado, and Ellie insisted on keeping the vents closed. The installation of screens will eliminate the bugs and allow the air to flow again. For smaller insects the Van Winkles have an organic means of extermination: a family of garter snakes living in the rocks below the windows was trapped inside when the greenhouse was built.

Greenhouse Storage

Since a solar greenhouse performs better with storage capability, Rip set out to find something that would provide all it would need. A 2,400-

Rip explains chart on fuel consumption.

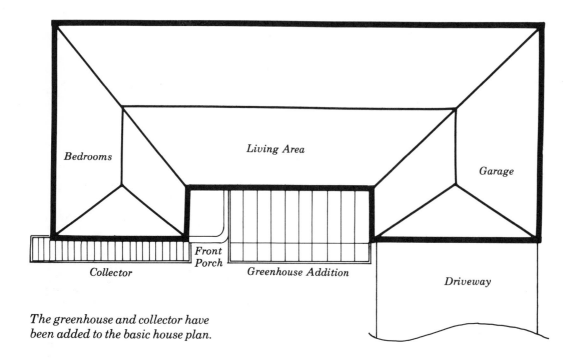

The greenhouse and collector have been added to the basic house plan.

gallon cistern is sunk in the ground beneath the greenhouse and insulated with an inch of polystyrene bead board on the outside and another inch on the inside. "I should have used twice as much insulation," Rip says. The purpose of the large volume is to "minimize storage temperature swings in order to provide maximum Delta T between the tank and the collector," Rip explains in his engineer's language. In other words, the higher the differential between the storage tank and the collector, the higher the efficiency.

It didn't work out as anticipated. The water in the storage tank never got above 68°, meaning that at night greenhouse temperatures descended into the 30s. When the storage medium is the same as the transfer medium, designing too large a storage area is a common mistake made by solar builders who are afraid they'll be unable to trap all the solar energy received. Various solutions were possible for Rip, such as extending the collector area or reducing the size of the storage by partitioning it. Rip's answer was to attach a matching active collector at the other end of the house to heat the water to the desired temperature. He had room for enough collector under the bedroom windows to provide space heating too.

Cross-section of greenhouse.

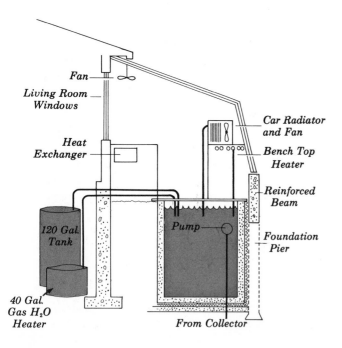

Distributing the Heat

He divided the tank into compartments with Styrofoam partitions coated and bonded to the tank with masonry stucco. One holds 1,500 gallons; the other two, 350 gallons each. Water is drawn from the coldest tank and circulated through the active collector. A differential thermostat makes sure that water warmed in the collector is returned to the hottest section that can still accept an increase in temperature, thus improving the efficiency of the system. When the greenhouse temperature drops to 50° F., water is taken from the warmest tank and pumped to a heat exchanger beneath a planting bench. To provide more heat, water is sent at the same time to an old Ford automobile air conditioner with an attached fan. Since these modifications, the greenhouse has maintained a nighttime temperature of between 40° and 50°.

Did Work Themselves

"There was a lot of fun in the whole project," Ellie says, "but it was a lot of work." It took six solid weeks to build the greenhouse and five to put in the active system. Ellie, a botany teacher, now spends up to ten hours a week in the greenhouse weeding, watering, pollinating. Some of Rip's time is taken up every day in keeping accurate records.

"The Ripper," as Ellie calls him, did all the soldering on the active collector, spending a week on it. He recommends that others do not undertake this task themselves. Asked why he used copper plate instead of cheaper aluminum, Rip replies: "Copper has six times the conductivity of steel and twice that of aluminum. Besides it's easier to work with, which is important for the owner-builder."

Ellie and Rip working in the greenhouse.

At the time of its purchase, copper cost $2 a pound. With each pound providing 1½ square feet of plate, Rip invested about $1,000 in the absorber. An elec-

Schematic of Van Winkle solar system.

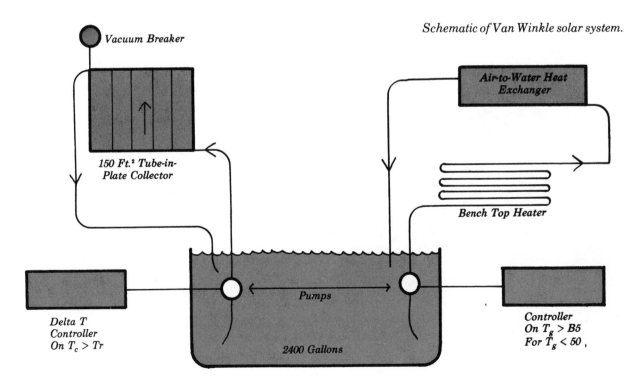

trical engineer for IBM, he had the know-how to make a control system using solid-state circuitry for $100. Technically it was not a computer. Rip calls it a "decision-maker."

Troubles

Since its completion, the active collector has posed several challenges. Because of the length of the collector, the water flowed faster at the ends than in the middle. Restrictions had to be put into the pipe to even up the flow. Getting the solder water-tight was next to impossible. "I've got some leaks," Rip admits, "and the insulation on the back doesn't fit as tightly as it should. But we still get 100,000 Btu a day." The glazing is either too tight or too loose, depending on how you look at it. It's just perfect enough to let in fine particle dust at the top, but has no holes big enough to permit the use of a vacuum cleaner to get it out. "Better not try to make your collector airtight unless you can make it **absolutely** airtight," Rip warns.

Rip explains data to solar tour group. His home-made sound system includes a radio transmitter hooked into his stereo speakers.

Rip keeps accurate records of the performance of his system.

What with the dust, wild morning glories have sneaked into the collector underneath the glass at the bottom, stealing some of the solar heat. Still, water in the copper tubes is at 150° F., the tour group notes. "Have you had any problem with hail?" someone asks. "We've been real lucky— we've had hail the size of mothballs but no glass has broken," is the reply. With wind out of the northwest the collector is protected. "I don't know if it could take peach-sized stones though."

By heating the basement, the active collector provides about a quarter of the Van Winkles' space heating needs, a little less than the greenhouse. Warm air from the greenhouse is admitted to the house through casement windows in the living room and from there to the dining room, kitchen, and utility room—an area of 1,200 square feet. Besides going to the greenhouse tank, water from the collector is run into a 120-gallon storage tank/ heat exchanger to provide space heating for the

The Van Winkle living room. Ellie lets it heat up to 85° during the day in winter.

basement. In all, Rip estimates that half their heat comes from the sun, while nearly all their domestic hot water is provided by the heat exchanger in the storage tank.

Solar Living—Costs and Benefits

So what if it's not a perfect system. It works well enough and it was cheap. "It makes our pocketbook feel just great," says Ellie. The Van Winkles are paying a fourth what they paid for gas in 1966 even though rates have risen. The cost of natural gas in Boulder is still so low that saving on fuel is not much of an incentive for their neighbors to switch to solar heat. If the alternative to solar is electric resistance heating, however, the incentive is three times greater.

Ellie found that it takes a while to get used to living in a solar home. "You have to learn when to close and open vents, and remember to put on clothes instead of turning up the thermostat when your husband won't let you. Nor can you do too much wash at one time."

She feels secure, however, in the knowledge that if the electricity goes off during a storm they will stay warm: "You always have a warm place to go. Maybe you have to huddle in the greenhouse, but it's better than freezing."

The whole project began because Ellie wanted a greenhouse and a solar-heated swimming pool. She delights in the greenhouse but despairs of ever getting the pool. "It'll never happen unless I get out there with a shovel and dig it myself." The pool is just too big a project, requiring a 300- or 400-square-foot collector. Before Rip can tackle that, he'll have to put in a wider sidewalk to salvage the lawn destroyed by solar tours.

Most Outrageous Solar Greenhouse

Never put off until tomorrow what should have been done in the early Seventies.

George Ade, 1901

No doubt this is one of the first solar homes in the country to be into its second owners. Why buy a secondhand solar home? Well for one thing you know what you're getting. Sara and Douglas Balcomb purchased this home in 1977, after the doctor for whom it was custom-built moved his practice. Doug is senior scientist in the Solar Division of the University of California Scientific Laboratory at Los Alamos, which had been monitoring the house. Because he was familiar with the results of the performance analysis and appreciated the esthetics of the house, when it came on the market he jumped at the chance to buy it. Are the Balcombs satisfied with their investment? "Why, sometimes

I think nobody has the right to enjoy this much luxury," Sara says. "It's outrageous, almost indecent, to be sitting in a greenhouse during a snowstorm, among lettuce plants and orchids."

The house is located in First Village, a small, planned environmental commmunity six miles south of Santa Fe. All the homes in the village are equipped with solar-heating systems, ranging from an active air installation to a water-filled Trombe wall. This, the first to be built, is a two-story structure emphasizing passive-gain principles for space heating, along with an active system for domestic hot water.

The Balcomb home wraps around a large, triangular-shaped, 2-foot-high greenhouse. The north side of the house, sunk 4½ feet in the ground, is curved to offer less surface area to winter's chill winds. While the south wall of the greenhouse is entirely glass, two interior walls constructed of adobe bricks set with mud mortar afford thermal mass and separate the heat-collection area from the main living spaces.

When the winter sun, entering at a low angle, heats up the adobe, the greenhouse side of the wall may reach 120° F., while the house side, after a time lag of 6 to 10 hours, gets as hot as 80° F. After the wall attains its maximum, it loses heat slowly, continuing to radiate warmth all night.

With the exception of the baths, all the rooms share some of the heat from the thermal walls. Electric heaters in the bathrooms are sometimes used during the early morning or evening hours. Fortunately you needn't go through the greenhouse to get to the baths, a fact that is appreciated at 3 A.M. when it's 50° F. in there.

Live-in Greenhouse

Temperatures in the greenhouse vary widely as there is no mechanism for retaining nighttime heat or providing shade. The coldest last winter was 45°, while 58° was the average. The Balcombs find that most plants can tolerate such fluctuations: tomatoes don't do well, but leafy vegetables such as lettuce thrive.

In the daytime even when snow is falling outside, the greenhouse temperature always reaches 65°, so it is used as a solarium. In the summer the sunroom hits the 90s, although never exceeding the outside temperature. Ventilation is provided by vents along the bottom of the south wall complemented by a large window vent at the top of the stairway. A strong convection is established, circulating air up the stairwell and out the manually operated vent to cool the whole house. In addition, the adobe wall is shaded by a balcony and a roof

---

TECHNICAL DATA

Owners: Douglas and Sara Balcomb, New Mexico
Designer: William Lumpkins (Sun Mountain Designs)
Builders: Susan and Wayne Nichols (Communico)

General Features
Latitude: 36° N
Degree-days: 6,000
Insolation: 250
Heated area: 1,950 ft^2 plus greenhouse
Year of completion: 1976
Insulation: Walls: 2" polystyrene below grade; 7½" fiberglass
     above
   Roof: 4" polystyrene
   Foundation: 2" polystyrene
Orientation: S
Solar system: Hybrid—greenhouse with fan

Collection System
Passive collector: 400 ft^2 greenhouse with double glazing
Angle: 60°
Storage/distribution: 10"–21" adobe thermal walls radiate
  heat into living areas
Active collector: Two fans take air off top of greenhouse
  through ducts to rock storage. Each fan is ⅓ horsepower,
  moves air at 1,100 cfm.

Storage System
Material: 24 cubic yards of rocks
Location: Under floors of living and dining rooms
  and kitchen

Distribution System
Heat trapped in rock beds conducts up through 6" concrete slab and quarry tile floor and radiates into rooms.

Auxiliary System
Backup: Electric baseboard and fireplace
Fuel consumed: $38 in electricity 1977-78; approximately ½ cord wood. Total electric usage for heating only for 12 months was 857 kwh.

Domestic Hot Water
Collectors: Two 4×8 panels in backyard
Manufacturer: Miramet (Israel), distributed by American
  Heliothermal
Angle: 60°
Cover: Single layer of glass
Absorber: Copper pipes attached behind steel plate
Storage: 40-gal. electric water heater

Costs
House: $69,000
Solar space heating: $8,300 (including greenhouse with
  adobe wall)
Solar hot water: $5,200
Special funding: Two-thirds of cost of solar offset by HUD
  demonstration grant

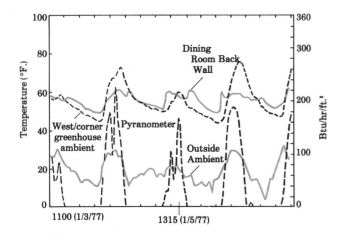

Representative temperature data, January 3-7, 1977.

overhang. During the blistering days of July and August in 1977, the wall temperature never went above 76° even when it was 97° in the front yard. Because the house is sealed off from the greenhouse by double-glazed doors, the temperature inside rarely varies more than five degrees from an average of 62° F. The greenhouse also serves as a buffer zone between the rest of the house and the elements, replacing an air lock entranceway and a boot room. As the only furniture is lawn chairs, fading is not a problem.

Additional Heat Storage

Since more heat is trapped by the greenhouse than can be stored in the adobe walls, two additional storage areas are charged by air drawn off at the roof peak. Fans draw the air through ducts between the first and second story and under the living room and dining area. Up to 10 million Btu can be stored in the rocks, enough to carry the house through two or three sunless days.

After three cloudy days, when the rock beds have reached ground temperature, the fans come on earlier and run longer, restoring the floor temperature in one day of sun. The controls are simple: a differential thermostat activates the blowers while a backdraft damper prevents reverse flow. The fans, which operate about four hours each day, are so quiet that the only background noise we hear emanates from fans that prevent overheating in the computers installed by Los Alamos Laboratory.

The rock beds are a low-temperature storage medium, only heating to 75° at the maximum. "The floors never feel warm to the touch," says Sara, "but you really notice the difference if you walk barefooted from the living area into the utility room, which has no storage underneath." Actually the rock beds do not charge; rather there is a continued upward flow of heat. The stones are 4- to 6-inch round river rocks, somewhat larger than are usually found in storage bins. According to research Doug conducted at Los Alamos, size doesn't matter much if the rocks are all fairly uniform, so that little pebbles don't fill up the spaces and impede airflow. Sizing the fan in relation to the storage is also important for achieving maximum efficiency. In this case two 1/3-hp fans move 1,000 cubic feet of air per minute (cfm). A little experimenting may be necessary to obtain the right setup. The Balcombs' fans could be used to reverse the airflow for summer cooling, but New Mexico nights make this unnecessary.

Most Successful Greenhouse

We were first inspired to visit the Balcomb home by Bill Yanda, an expert who has built hundreds of solar greenhouses. Yanda calls it "the most successful greenhouse-heated solar home in the country." It is almost 100 percent solar-heated—last winter the total fuel bill came to $38. An electric baseboard register in the living room was used so infrequently that it was replaced with a bookshelf.

Schematic of Balcomb heating system.

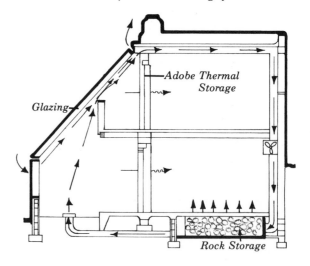

Due to a freak storm, the toughest test of the system came in May, of all times. A four-day blizzard dumped 3 feet of snow on Santa Fe. The house was chilly and the backup came on. "Solar houses aren't made for that," says Sara. "They're made to be tested in January, when the sun is at the right angle."

Even during difficult periods, however, the house consumes little electric energy. The utilities claim that solar-heated houses are going to require more electricity during peak hours, because people will turn up their thermostats when the sun goes down. But in the Balcomb house the solar heating doesn't "bottom out" until after midnight.

Comfort

The other attraction of the house is its livability. Featuring three bedrooms on the second floor opening onto a balcony inside the greenhouse, it

View of front entrance from second-floor balcony, which provides access to bedrooms.

The Balcombs' greenhouse in summer profusion. The 14-inch adobe walls are shaded in summer. In winter heat stored in them is radiated to the interior rooms after a time lag of 10 hours.

has touches of elegance as in the solid pine stairway and gingerbread balcony banister. Downstairs, the breakfast nook is intended to catch the early morning sun, while the living room is conveniently located in the southwest corner. Due to the passive heating, Sara finds the house exceptionally comfortable, with no hot air blowing around, no static electricity, no drafts, which means no sore throats, no dry skin, no furniture cracking and splitting. "It's so stable it's almost boring."

Furthermore the house was designed for real human beings. When interviewed by people from the Department of Housing and Urban Development, who complained that early HUD-sponsored houses hadn't sold well, Sara asked what they expected. "The engineers looked at the roof bracings and the architects looked at the lines and nobody noticed how far it was from the front door

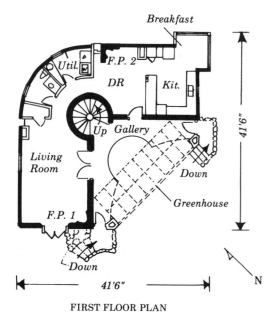

FIRST FLOOR PLAN

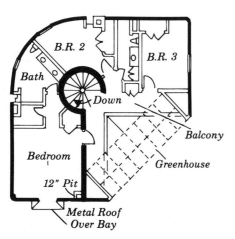

SECOND FLOOR PLAN

housewife and to the construction industry. She sees her task as a destroyer of myths: you don't have to wear sweaters; and you don't have to change your life-style completely. "Our house is quite comfortable in winter even when I'm not wearing a sweater," she insists. "And I spend little time hassling with the solar heating, apart from five hours a week in the greenhouse, if you want to count that. The plants are phototropic—they follow the sun—and you have to turn them to keep them growing straight. Using plastic on the greenhouse or painting the rear wall a light color avoids this chore."

Sara theorizes that contractors who aren't building solar houses are playing it safe because the construction industry nearly died four years ago. "Some contractors practically had to send their daughters out to walk the streets." As officers of the New Mexico Solar Energy Association, the Balcombs have worked to change building codes and the training of inspectors to improve the climate for solar energy. In nearly every state the

Because of the north-facing windows, the breakfast nook requires auxiliary heating.

to the kitchen." Unlike this one, many of the first houses were too open, affording no privacy: "They'll appeal to a certain number of people, but when you're a mother with two screaming children, you need someplace to nurse your headache." That's why it's important to have a woman in a construction firm. Sara attributes much of the success of her home to the genius of builder Susan Nichols: "She knows how to avoid making the little mistakes that can turn off people who aren't deeply committed to solar heating."

Destroying Myths

As an enthusiastic proponent of solar energy, Sara would like to take her message to the American

*The Balcomb house is shaded
in summer afternoons by
juniper trees.*

codes need revising to encourage the use of energy-saving methods while allowing a person to build the kind of home he wants and needs. New Mexico has now instituted a mandatory training program for building inspectors to insure that they understand the latest technology.

"Passive heating may be 'low tech,' but it's not simple," says Sara. "It's sophisticated beyond belief. When you start to design a home, you really have to sit down and do your heat load calculations." Often people fail to think ahead. They build a Trombe wall and wonder why they can't see out, or a direct gain system and wonder why fabrics fade.

Value of Rock Beds

In the Balcomb house the big question is whether the rock beds are useful or could have been omitted in favor of a passive system. The house has been monitored by Los Alamos since 1976, with a weather station on the roof as well as an hourly check of thirty-nine data points and a continuous audit of electrical usage. The statistics indicate that the rock beds are worth the extra cost, both in terms of preventing overheating in the greenhouse and increasing comfort with warmer floors. Doug's conclusion: "The rock beds add up to a quarter of our

heat, which would otherwise be supplied by the electric backup, and only cost about $1,000, including the insulation and ductwork."

Quick Financing

The Balcombs' experience with financing reveals an interesting situation. The lowest interest rate they found at a bank was through a guaranteed federal loan program, but the red tape would have taken three months. Instead, the local savings and loan gave them a preferred rate mortgage in one day. Sara feels that the federal government should speed up the process of making loans for solar projects. Despite this encounter she is optimistic about the future of solar heating in the United States, having seen the progress made by the New Mexico state government.

Doug Balcomb has written many useful publications on passive solar heating. Summaries of his findings have been published in the *Bulletin of the New Mexico Solar Energy Association,* P.O. Box 2004, Sante Fe, N.M., 87501. In addition, he is a member of the board of directors of the International Solar Energy Society, which, with 700 members in 55 nations, promotes solar research, provides public education programs, and fosters application of available information.

The House That Sarah Built

What will not woman, gentle woman dare...

Robert Southey

With just a trace of indignation Sarah complains that everyone says, "I hear Chris and Jason built your house." They never say, "I hear that **you** and Chris and Jason built your house." "They must think I just sat around making coffee." Quite the contrary. As the third member of the crew, Sarah Sanford worked on almost all phases of the construction. Although she makes her living as a nurse in a small hospital, Sarah is qualified to pound nails. Back home in Vancouver, British Columbia, she spent a year learning carpentry in a

cabinet-making cooperative before trading one ocean for another by moving to the coast of Maine.

Once in Maine, an inheritance enabled her to acquire a 25-acre hillside covered with spruce and blueberries. A friend, Stevan Grabara, designed a contemporary saltbox that now stands as a sentinel facing the sea like many of its predecessors of the 1700s. Stevan supervised the construction while Sarah swung a hammer and handled the bookkeeping.

Thoughtful touches throughout the house reflect the same kind of care that down east ship captains lavished on their homes: handcrafted oak cabinets in the kitchen, hand-turned wooden railings for a stairway nearly as steep as a ship's

Domestic hot water collector on roof. The house has few windows on the east and west, and only one window and one door on the north side. (Courtesy Paul Bolduc)

ladder, a multipaned glass door leading from Sarah's bedroom to the bath, and a bookshelf door that opens like the entrance to a secret passage.

The floor plan resembles that of the traditional saltbox with the front entry leading to a central hall containing a stairway, but in this case a utility room branches off one side and a guest room off the other, with the rear of the house devoted to a living-dining-kitchen area. Sliding glass doors at each end of the big room flank the solar collector. Upstairs, Sarah's bedroom and a studio for weaving, her hobby, each have a glass door opening to a south-facing balcony while a bath and sauna are tucked under the steeply sloping north roof.

The simplicity of this design appeals to Sarah. "There's something in me that wants things simple and natural, as close to this good earth as possible. It's not just that I see an oil crisis coming. It's that I appreciate this water and those trees and that sunshine, and I love soaking it up in every way, visually and physically." The environmental motivation for building a solar home is stronger for Sarah than a concern with saving money. She describes herself as one of those people who divide the garbage into six bags. A vegetarian, she worries about what to feed her meat-loving cats; a compulsive electricity miser, she goes around the hospital at 3 A.M. switching off lights.

$135 Heat Bill

Not that financial savings are unappreciated. Her first year's heat bill ws $135, although Sarah believes it would hit $200 if she spent more time at home. In the morning she builds a small fire and leaves for work, relying on the sun to heat the house on sunny days but simply letting it cool down on cloudy days. In the evening she lights another fire, then lets it go out overnight. If the house drops below 58° F., the baseboard heat kicks on.

Credit for the low fuel bill goes to the design of the house together with passive solar heat, since the active air collector was unfinished until the last month of this winter. Stevan Grabara attributes the performance of the house to its compactness, insulation, and southern orientation, as well as a spruce windbreak. Howling north winds slide up and over the long saltbox roof rather than slam-

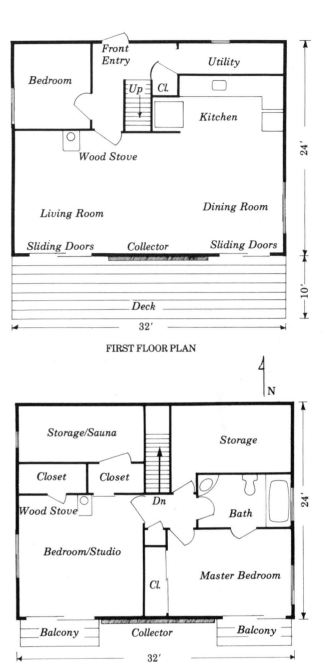

FIRST FLOOR PLAN

SECOND FLOOR PLAN

ming into a vertical wall. Most of all, according to Steve, there are no drafts. The exterior shell is not particularly tight since he was concerned about adequately ventilating the insulation, but the interior has a double vapor barrier: foil plus plastic sheeting.

Despite the success of Sarah's house, both she and Steve express reservations about the concrete floor. The slab was covered with quarry tile, esthetically pleasing and easy to care for, yet the floor feels chilly since there are only two sliding doors on the ground level for passively gathering solar heat. To warm that much mass you'd have to turn the entire south wall into an active collector—making the house dark—or use a more efficient, but more expensive, collector. Additional windows would do the job, but Steve was reluctant to add more passive unless the house had shutters. He argues that having movable insulation means that when you leave for a few days you're faced with a hard decision: do you leave it open or closed? If closed, you lose heat on sunny days; if open, you lose at night and on cloudy days. For this reason he wanted the passive system to be supplemented by an active collector in which a thermostat makes the decisions. But since most people travel infrequently, this hardly seems like a compelling argument for using an active system or for deciding against shutters. For one thing shutters can be automated, although the expense scarcely seems justifiable. Just go ahead and leave them open, or closed.

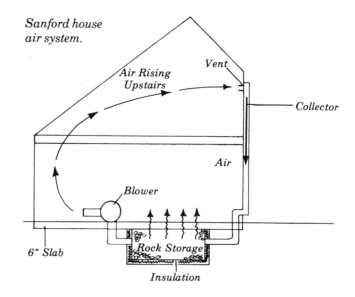

Sanford house air system.

Even more than adding windows or a larger collector, insulating beneath the slab would have helped overcome the problem of a cold floor. In a tight New England house an uninsulated slab can

TECHNICAL DATA

Owner-builder: Sarah Sanford, Maine
Designer: Stevan Grabara

General Features

Latitude: 45° N
Degree-days: 8,200
Insolation: 140
Heated area: 1,312 ft²
Year of completion: 1977
Insulation: Walls: 6" fiberglass
      Roof: 9"–12" fiberglass
      Foundation: 4"–6" Styrofoam on perimeter
Orientation: S
Solar system: Hybrid—passive direct-gain and active air

Collection System

Passive collector: 158 ft² (double-glazed windows)
Active collector: 192 ft²
  Angle: 90°
  Cover: Single layer 0.040" Kalwall Sun-Lite ®
  Absorber: Corrugated aluminum with corrugations running horizontally, nonselective coating, 6" fiberglass

Storage System

Container: 6" concrete slab and 9'×9'×16" (108 ft³) rock bin
Material: ¾" diameter rocks
Location: Bin beneath living room part of slab
Insulation: 2"–4" Styrofoam (sides and bottom)

Distribution System

Triggered by differential thermostat, ¼-hp blower draws air through two intake vents in upstairs bedrooms, down space behind absorber into two 8" ducts to rock bin. Air drifts through bin to a 10" duct to a utility closet containing the blower, then is exhausted through a register into the hall to return by natural convection to the intakes.

Auxiliary System

Backup: 55,000-Btu wood stove and electric baseboard
Fuel consumed: 1½ cords at $50/cord and 1,500 kwh at 4¢/kwh average

Domestic Hot Water

Collector: 32 ft² drain-down
Manufacturer: OEM Products, Inc., Brandon, Florida
Angle: 45°
Cover: 3/16" tempered glass
Absorber: Copper tubing on aluminum sheet, selective coating
Pump: 1/20 hp
Storage: Drains directly into 55-gal. electric water heater

Costs

House: $39,000
Solar: $1,200
Domestic hot water: $1,000

View of Maine Coastal islands from Sarah's bedroom.

account for up to a third of the seasonal heat loss. On the other hand, having the ground directly in contact with the concrete helps modulate house temperatures, the additional thermal mass reducing high-low swings. The latest thinking is that it may be best to insulate beneath a 2-foot bed of earth and then pour a slab.

If he were doing another house, Steve would probably eliminate concrete in favor of wood over rock storage. He might even forgo the rocks and vent the collector directly into the house. With vents at the ceiling and floor the heated air would

Sarah Sanford keeps her house fairly cool. She puts on a sweater instead of upping the thermostat or building a fire.

circulate by natural thermosiphoning. Such a system would lack storage, but Steve feels that pouring money into storage large enough for real carry-over in a severe climate may be a mistake when so much can be accomplished simply with proper insulation, good siting, and careful building techniques. He argues for putting money into an efficient wood stove and electric backup in preference to a complex storage setup.

Although the Sanford house is exciting proof of how much can be done by building a tight house, we can't see wholly eliminating storage. Making it big enough to guarantee all your heat needs may be uneconomical, but relying on solar gain on sunny days plus electricity and wood ignores the fact that electricity is exhorbitantly expensive and inefficient while the price of wood could start soaring as demand increases. As a matter of fact, storage is relatively inexpensive. Including enough to provide at least 50 to 75 percent of the heat load in a northern home makes plenty of economic sense.

The Sanford collector, although beautifully simple in design, has one flaw: the ducts and registers are rather small for moving the volume of air required by this kind of system. Since the blower must fight the resultant resistance to airflow, 1/8 horsepower proved to be inadequate. Steve recommends using a variable speed blower to allow the possibility of making adjustments.

Corrugated Absorber

Construction would have been simpler, he also suggests, if the absorber had been flat rather than corrugated. Corrugated aluminum was used to increase air turbulence. Like water in a stream, air tends to flow faster in the middle because of friction along the edges. A layer of relatively still air is formed next to the collector, preventing the moving air from coming in contact with the absorber and thus reducing the heat transfer efficiency. Baffling of some sort is needed to agitate the air to get as much of it in contact with the hot absorber as possible.

According to Steve, recent studies suggest that a flat surface may accomplish this as well as one that is rough if the back airspace is decreased and the airflow slowed down. Although research on air collectors has lagged behind that done on liquid systems, a few technological questions like this one remain unresolved for both kinds of collectors.

At any rate, Steve had to use ten tubes of silicone caulk to seal the corrugated edge to prevent an interchange of air between the front and back sides of the absorber.

Hot Water System

Sarah took complete responsibility for choosing the solar-heated hot water system. She advises prospective solar homeowners to do their homework, keep an open mind, and be prepared for a little floundering around. Locating a supplier by consulting a listing in Donald Watson's *Designing and Building a Solar House,* she selected a self-draining collector rather than a system using antifreeze on the grounds that the heat exchange coil required with antifreeze supposedly reduces efficiency by 15 percent. Many people opt for it anyway, fearing that a drain-down system is risky in a cold climate: the pipes could freeze if the electric solenoid that controls draining were to malfunction. Sarah was told by the manufacturer that the

Vertical air collector made of corrugated aluminum with a fiberglass cover. Four sliding glass doors contribute passive solar heat. (Courtesy Claude Bolduc)

system had been tested and found to function satisfactorily in cold climates. If a power outage should occur, a valve would open automatically to drain the panels.

When the OEM collector was delivered, all 600 pounds of it, Sarah realized that she was going to have a problem getting it on the roof. It was too heavy to haul up a ladder, so it lay on the ground for two months while she sought a solution. One idea was to use a logging company pulp loader, but the final choice was a septic tank truck with an arm long enough to reach the roof.

That problem was solved. Next the plumber and carpenters refused to be responsible for attaching the collector, fearing that it would blow off in a sou'easter. So it sat up there for a few weeks tied by a rope. One windy morning as Sarah left for work, she looked back and saw it had slid across the roof and was dangling over the edge. She and a neighbor managed to shove it back into position, and then bolted it through the roof into the attic, using one-inch plywood under each bolt. "If that thing blows off," she says, "the whole roof will come off."

The collector was not up in time to evaluate its winter performance, but in summer Sarah finds that a day of sun gives her three days of hot water

Dining area. Tile floor over concrete slab soaks up heat from two sliding glass doors. Two more upstairs help gather solar heat, enough to heat the house on sunny days in winter with no help from fossil fuels.

Blower and flexible duct coming from rock bin beneath living room floor. The blower is in a small utility closet in a hall beneath the stairway. Hot air is vented through a register into the hall. A sensor on the surface of the collector and another in the livingroom floor kick on the blower when the surface of the collector is 3 degrees warmer than the floor. (Courtesy Paul Bolduc)

unless she takes a bath, which reduces it to two days. This seems about right for a collector that's half the size a family would need, but at $1,000 for 32 square feet it's an expensive system. Sarah has turned the electric water heater off for the summer and is finding that she tends to take her bath on sunny days—"you get into the rhythm of it."

Living in a solar home has made no other changes in Sarah's life-style, since she already was a conscientious energy saver. But Steve feels that some drastic changes are overdue in this country: "I've been over to India and Viet Nam: we all live like kings here." He feels, however, that people in this country can and will absorb much higher fuel prices before they are willing to make real sacrifices. The biggest change Steve would like to see is the disappearance of the single-family dwelling. "When people are lumped together, it's a lot easier to keep them warm. That should be the first step, but I think it'll be the last." He would also like to see us "think small" when building any kind of home. We don't need nearly as much space as we think we do. While in Calcutta, he overheard an Indian remarking about a building that it would hold 400 Americans—or 800 Indians. Our houses, like our cars, are going to have to shrink.

A Package Deal

If this were war, we'd have solar energy in a year.

William Heronemous

When their 16-year-old daughter convinced them to build a solar home, Pat and Don Ritter opted for the convenience of buying a home and collector together in a "prefab" package. The same truck delivered the collector components and house sections to the Ritters' building site, a lot in a Chicago suburb. The pieces, manufactured by Acorn Structures, Inc., were then put together by a contractor.

For the past few years the Ritter family had been living a yoyo-like existence. Don, an executive with a marketing research company, was transferred from Illinois to Massachusetts, then back to Illinois. When it came time for the most

Pre-cut home with liquid-type solar collector. Corrugated fiberglass on collector adds structural integrity.

recent move, Linda—the 16-year-old—said, "I've got a neat idea." The other five children, ranging from Amy, 7, to Mike, 21, picked it up enthusiastically and drew their mother into the solar campaign. Only Don remained sceptical—not unreasonably considering he was taking on the responsibilities of a new position, moving a family, and facing the prospect of having three children in college within a year.

Pat discovered Acorn by accident. A friend mentioned that her teenager had written a report for science class on solar energy. The research material happened to include a brochure from Acorn. Pat, who has since become an Acorn sales representative, felt that the company was "honest and straightforward" from the minute she and Don walked in. A salesman warned against building a solar house for purely financial reasons but said if they were also interested in pioneering and conserving energy, a solar home would make sense.

Although other companies are offering solar home packages with collectors supplied by outside sources, Acorn claims to be the only home manufacturer to design its own system. Not long after putting it on the market, they sold thirty-five to customers from the East coast to the Mississippi River.

Simple Collector

For a liquid collector, Acorn's Sunwave® is relatively simple. The literature describes it as "down-to-earth, not space-age" technology that is already competitive with electricity, although not yet with oil. The current list price is $6,500 and installation averages another $1,400. Monitored for four years on a 1,400-square-foot model house, the collector provided 46 percent of the heat load.

In addition to custom-designing to your specifications, the company offers two standard Cape Cod models particularly suitable for solar collectors. Our Yankee ancestors, facing the yearly chore of getting in a wood supply, had energy conservation in mind when they dreamed up the trim, compact Cape. Acorn's version has a cathedral ceiling in the dining area, but potentially wasted heat is recirculated through a return air register near the roof. Windows are double glazed and concentrated on the south side; corridors and bedrooms act as buffers on the north. It's curious, though, that the Ritter house is inadequately insulated. Today's standards for a well-insulated home are double what Acorn put in Don and Pat's house. With ash doors, oak cabinets, and hardwood floors, the Capes are expensive, but Acorn does offer cheaper alternatives, including a solar garage for older homes unsuited to retrofitting. The solar-heated water is piped underground from garage to house.

Stephen Santoro, Acorn's chief solar engineer, says he is a firm believer in keeping things simple.

TECHNICAL DATA

Owners: Pat and Don Ritter, Illinois
Designer: Acorn Structures, Inc.

General Features
Latitude: 42° N
Degree-days: 6,639
Insolation: 140
Heated area: 2,600 ft^2
Year of completion: 1977
Insulation: Walls: 3½" fiberglass
          Roof: 6" fiberglass
          Foundation: 1" Styrofoam
Orientation: S
Solar system: Active liquid drain-down

Collection System
Collector: 418 ft^2 net
  Manufacturer: Acorn Structures, Inc., Concord, MA
  Angle: 47°
  Cover: Single layer Filon® (polyester fiberglass)
  Absorber: Copper tubing with attached .020" aluminum fin-plate, nonselective coating, 2" fiberglass
  Pump: 1/3 hp

Storage System
Container: 8'4" diameter×7' high cylindrical tank (⅜" pre-curved plywood reinforced with galvanized steel bands and lined with 30-mil vinyl)

Material: 2,000 gal. water containing algaecide
Location: Basement
Insulation: 3½" fiberglass (top and sides), 1" Styrofoam and 3" loose-fill vermiculite (base)

Distribution System
Two-stage thermostat turns on 1/12-hp pump, which sends water from tank through heat exchanger in duct. Auxiliary heater's fan blows warmed air to house. If water is not warm enough to satisfy thermostat, house cools until second-stage contact closes, turning on backup furnace.

Auxiliary System
Backup: 175,000-Btu gas furnace
Fuel consumed: $520 at 21¢ per 100 ft^3 (Sept. to mid-March, 1977-78, while backup was being adjusted)

Domestic Hot Water
30-gal. stainless steel tank submerged in storage tank pre-heats for conventional water heater.

Costs
House: $98,000
Solar: $7,200

But why choose a liquid system if you're after simplicity? Steve cites the electricity consumed by the added fan power needed in air systems and the questionable cleanliness of rocks: because solar heating involves large temperature fluctuations, where rocks can become cool while the collector is quite hot, Acorn believes that condensation will form in the storage, possibly creating mold and mildew. When the blower turns on, it would distribute this to the house. We cannot report, however, that any owners of air systems complained of this problem.

Acorn's collector, a closed system with the water contained in tubes, is conventional except for one feature. The fiberglass cover, Sunwave, has corrugations that add structural strength and take up thermal expansion. Acorn chose Filon because the company found it capable of retaining a wave while expanding and contracting. The added structural integrity seems to prevent flapping in high winds; the Ritters find their collector almost noiseless. Filon should not be used in an open system like a trickle collector since tests by Brookhaven National Laboratory have disclosed

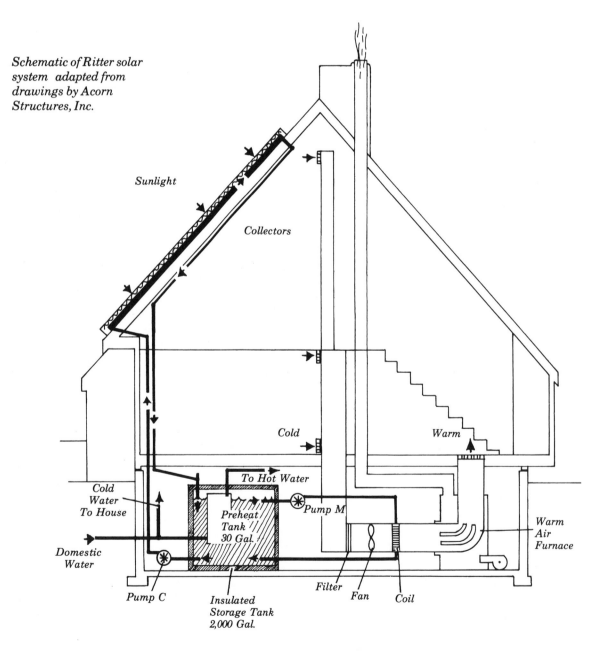

Schematic of Ritter solar system adapted from drawings by Acorn Structures, Inc.

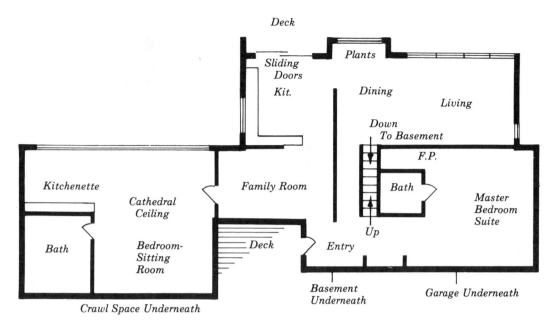

FIRST FLOOR PLAN

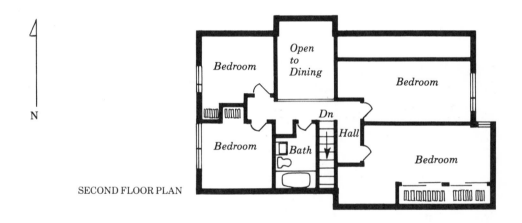

SECOND FLOOR PLAN

that Filon deteriorates rapidly when exposed to hot water.

The storage system also has one distinctive feature: the tank is wood lined with plastic. Acorn has thus far had no problems with leaking tanks. They *have* had two cases of collector leaks, both caused by improper installation. In one instance Steve discovered that the contractor, convinced that he knew how to do it without instruction, ignored the installation manual.

Acorn conducts two inspections, one during construction and the other upon completion. In the beginning they had the individual trades take responsibility for the various phases—the sheet-metal worker putting in the heat exchanger, the plumber doing the plumbing, the electrician doing the wiring, and carpenters putting up the collectors. But Acorn has "changed its conception of how the whole thing comes together" and now persuades the heating contractor that it's worth his while to handle the entire process, installing both auxiliary and solar heating systems. Steve says that this eliminates most mistakes, but the "human element" remains one of Acorn's problems.

High Heating Bills

The Ritter house is a case in point. Pat is remarkably frank about the problems she and Don encountered. The first heating bills ran high: $30 a month in the fall. This was before the collector was hooked up. But both before and after it started operating, 62° F. was the warmest the main part of the house would get except when sunlight was directly warming the rooms through the windows. The Ritter residence is Acorn's 1,950-square-foot model with some revisions in the floor plan. A 650-square-foot extra wing contains a large bedroom-sitting area with kitchenette—a private apartment for extended visits by Pat's parents. The mid-50s was the top heat reading for that section.

The heating contractor immediately blamed the solar system, insisting that the temperature in the storage was too low to heat the house adequately. We can't resist commenting on the absurdity of this conclusion: if the collector is not providing enough heat, the backup must come on automatically after the house cools a degree or two below the thermostat setting. Nevertheless on December 1 the heating man disconnected the collector to determine whether it was causing the problem. He

The cylindrical wooden solar storage tank is in the basement under the main part of the house. The tank consists of six curved plywood panels steel-banded together.

hooked it up again on January 8, after a period of nearly six weeks during which the house remained just as cold. The obvious was proven: the culprit was the backup, not the solar.

During this time Chicago was hit by 2½ days of extremely cold weather. An adequately sized furnace operates only intermittently, but the Ritters', though it should be more than up to handling any heating demand, was forced to run continuously for sixty hours. The heat bill from mid-November to mid-January was $258. It was then discovered that the return duct going across the attic was uninsulated, even though 6 inches of fiberglass separate the attic from the rest of the house. With temperatures up there dropping as low as 25°, the furnace was receiving super-chilled air instead of normal return air.

Several degrees of temperature were gained by correcting this and other mistakes. Acorn recommends putting half the return registers near the floor and the other half on the wall. The Ritters' heating contractor put them all high. He also failed to install one in the vicinity of the thermostat, which is in the front hallway. This meant the thermostat was in a dead-air space. "We were stuck," says Pat, "in the situation where we had an existing system, and there were only so many options before beginning to tear out walls." It turned out to be feasible to install only four low returns, although fortunately one is near the thermostat.

Ducts Taped

The next improvement, suggested by Steve Santoro, was taping the ducts. Don and Pat did the work because by then the heating contractor was taking the attitude that his job was finished, despite not yet having provided the Ritters with a warm house. Taping her way out from the furnace in the basement to the crawl space under the wing, Pat could feel warm air blowing on her face and hands.

Next it was discovered that the builder had neglected to insulate the living room floor, located above an unheated garage where temperatures drop to 20° F. After rectifying this omission, the Ritters found they could get the main part of the house as warm as they wanted it. The addition, though warmer than it was, never climbs above the low 60s. The next fuel bill was a little better, $200 from mid-January to mid-March.

Cold Spots

When solar-heated air is coming out with no boost from the backup, the main part of the house still has cold spots near the vents closest to the furnace. The storage averages about 90° F., the same as Acorn's test house, with air exiting the registers in the low 70s. Combine rapidly moving air with low temperatures and you have discomfort: "I get out of bed in the morning and feel like I'm standing outside in a good strong breeze," Pat says. "I was in Concord three weeks ago and made a point of going into the model house, and unless I missed the duct where it's blowing like the dickens, you just don't have that sensation." She blames the heating man for simply not understanding that solar air, being low temperature, has to move slowly and constantly.

They could live with this from a comfort standpoint by donning extra sweaters, but what disturbs Pat is that they are forced to turn the thermostat up to 70° F. or above to compensate, because you can't keep a sweater near every cold spot, nor put on and remove clothes as you go from one area of the house to another.

So the Ritters are still searching for a solution to this problem. Meanwhile they had to buy an electric space heater to keep the apartment warm for Pat's parents. For the time being Pat says she's just "going to have to say nuts to the electric bill." The heater runs a lot but keeps the wing at 68° F. Interestingly that temperature is comfortable in the apartment where the air moves slowly, but not in the main house: "In here we freeze to death at 68°."

Five months of frustration and fighting with the heating contractor have not discouraged the Ritters, but they do have to keep reminding themselves that a pioneering project sometimes takes extra work. Pat says philosophically that there have to be some guinea pigs. You never find out in a laboratory situation how a real family does in a real house. She blames the heating contractor, not Acorn, for their troubles. He seemed to take the

Apartment wing with kitchenette. Pat believes they should install a blower with a separate thermostat to try to get more heat into the wing.

tongue-in-cheek attitude that the Ritters were simply playing with a new toy, a gadget, a conversation piece. Midwesterners, Pat says, seem sceptical of solar heating. She believes that if they were to build the same house in the East, they would encounter fewer problems.

Much of Acorn's help to the Ritters was *ex post facto,* after the system was installed. The Ritters' experience indicates that solar heating companies should minimize the human errors by providing careful supervision. When they don't know the contractor, frequent inspections are essential. Having a sales rep like Pat in a newly opened area should help.

Collector Sizes

An obvious explanation for some of the Ritters' problems is the size of the house relative to the size of their collector. Acorn disagrees with the rule of thumb that the collector area should equal a third

Pat Ritter, now a sales representative for Acorn Structures.

or half the square footage of the house. They put the same system on their 1,500- and 2,900-square-foot models, arguing that if you go by the rule of thumb, you end up with a technically appealing system but not a cost-effective one. Since the larger home has a higher heat loss, the solar gets called on more. This drives the temperature of the storage down, making the collector more efficient. The greater the temperature differential between the storage medium and the collector, the more efficient the heat transfer.

Let us take a hypothetical case to illustrate Acorn's argument. Suppose in a given climate the 1,500-square-foot house would run up an oil bill of $500 if it had no solar heating, and the 2,900-square-foot model had a bill of $1,000. Suppose Acorn's 400-square-foot collector provides 50 percent of the heat load of the smaller house and 30 percent for the larger home. The solar savings on the small house is $250 and is $300 on the large house. Thus on the 2,900-square-foot house the same collector supplies a lower percentage of the heat but a $50 greater savings. Since the installed cost is the same in both situations, the collector in the larger house is more cost effective.

This argument represents an oversimplification. Following Acorn's line of reasoning the smaller the collector, the greater its cost effectiveness so that a one-square-foot collector on a 10,000-square-foot building would be an ideal solution. Emphasizing cost effectiveness obscures the fact

that a proportionately larger collector on the 2,900-square-foot house would save $500, a net gain of $200, despite its lower technical efficiency. Given two systems that both save money, what the house buyer is looking for is not the one that saves money most efficiently but the one that saves the most money. The optimal solution has to be worked out in each situation; it depends on climate, volume of the house, house design (insulation, window location), design of the collector, storage capacity, and life-style of the occupants. This suggests the buyer should ask Acorn or any designer to evaluate the cost effectiveness of different-sized systems on the house he has chosen in order to make the best choice.

Solar Outlook

From his vantage point as an expert, solar engineer Steve Santoro sees the outlook for solar as changeable. Much depends on what the government does or fails to do. One after another, the major roadblocks to solar heating are being knocked over. The reluctance of banks to finance

North side of house, showing wing at left for apartment for visits by Pat's parents.

solar homes was the first to fall. The banks are now looking at solar heating as a good investment because it reduces the operating costs of a house, leaving tax-free savings which can be used to help pay off the mortgage.

Insurance companies provided another roadblock. In a two-day meeting with executives from major companies, Steve discovered that a Catch-22 situation existed. The insurance companies said they lacked the millions of statistics necessary to set up a policy rate structure for solar homes. The impasse: you can't write insurance policies until you build millions of solar homes, and you can't have millions of solar homes until you can get insurance for them. At the meeting, when the executives learned more about solar equipment—especially that it's not frighteningly complicated—they relented and agreed to insure based on the increased value of a solar home. If a standardized surcharge were added instead, a definite potential for customer rip-offs would exist.

Until recently the biggest roadblock was the federal government. It was once said of another Congress, but is appropriate here, that it doesn't run—it waltzes. As long as Carter's energy program remained a political football, the promise of tax credits for solar heaters persuaded some people to delay investing until they could receive a rebate. Now that federal tax credits of up to $2,200 are available, financial stumbling blocks are going. Now, if we can just deal with the human element: heating contractors who are set in their ways....

Stairway to the Sun

Nearly all the people of all ages involved in solar energy have a quality of the dreamer about them.

Daniel Behrman

In 1960 you could do a complete inventory of solar buildings on your fingers and toes. Those early buildings consisted mostly of test installations on university campuses, not real homes. So Norman and Jeanne Saunders were nearly two decades ahead of their time when they took out a loan to build a solar home for themselves and their four children. They obtained a mortgage with no trouble because the bankers thought Norman was just playing games with solar heating. "They simply ignored it—didn't take me seriously." Today people pay attention, particularly because the Saunderses were farsighted enough to see the importance of keeping solar costs low.

Maybe Norman's early advocacy of inexpensive solar systems reflects pure Scottish stinginess:

that's what Jeanne claims. In a more serious mood she talks of how he dreamed of finding a better way of life while minimizing damage to the environment. Norm and Jeanne were conservationists long before nature-loving became fashionable. A sadness creeps into Jeanne's normally energetic voice when she recounts the change overtaking their semirural Boston suburb: "This was 65 acres of woods and meadows when we first moved here. We had bluebirds, hermit thrushes, thrashers. Now the urban birds are infiltrating."

Jeanne works as an aide in the foreign language department of a secondary school, while Norm is a professional engineer who made his living in circuitry design until four years ago when he began devoting all his time to solar energy. His interest dates back to the mid-1940s when a *Popular Science* article on solar heating set him on the road of invention. Inventing runs in the family. His father was a salesman who built a refrigerator and electric mixer while everybody else was content with iceboxes and eggbeaters.

The Saunders house, which Norm named "Experimental Manor," was built from stock plans that he and Jeanne modified for solar heating. They used earth berms on the north, heavy roof insulation, double glazing, and large expanses of glass on the south. A concrete slab was poured on a metal sheet on top of concrete blocks. Pumice blocks were used for the east and west walls. Because of its spongelike airspaces, pumice serves both as a heat store and insulator, with a thermal resistance superior to that of ordinary concrete block although no better than an insulated stud wall. Since a pumice wall will lose heat more rapidly to the cold outdoor air than it will to the warmer room air, it is a marginally effective storage system and would be greatly improved by adding insulation on the outside. Today most designers using thermal mass do insulate on the exterior.

In spring and fall, the Saunderses find that heat carryover is about twenty-four hours; in midwinter about half that. Whenever the house heats up abnormally high, a large fan moves the warm air to storage beneath the slab. This fan was installed to pull air down from a series of rooftop collectors that Norm experimented with during the 1960s. The roof slopes 3° to the north, just enough for rain to run off, so with typical irony Norm boasts he had the only collectors in the country mounted on a north roof. The first was a simple sheet of polyethylene suspended over blackened aluminum. When the polyethylene deteriorated, Norm junked that system in favor of collectors in sawtooth rows. Being long and low, they were frequently snowed over.

New Collector

The collector that followed, built with the help of son Laurence, consisted of a thousand panes of glass in two layers parallel to the roof surface and supported by foam glass insulation. The pilot of a small plane once reported that from the air the house looked like a pond. When breakage occurred, Norm hypothesized that a duck may have tried to land. Some glass broke because too much elastic adhesive was used to seal the panes, leaving no room for expansion. The glass from this collector eventually was salvaged and used to build a sunroom by enclosing the area beneath a second-floor balcony. Norm says that sometime in the 1960s it dawned on him that probably he had

---

TECHNICAL DATA

Owner-designers: Norman and Jeanne Saunders, Massachusetts

General Features
Latitude: 42° 30' N
Degree-days: 6,000
Insolation: 115
Heated area: 2,625 ft^2
Year of completion: 1960
Insulation: Walls: Pumice blocks filled with exploded mica
(equivalent to stud wall with 3½" fiberglass)
Roof: Equivalent to 13" fiberglass
Orientation: S
Solar system: Hybrid—passive with fan

Collection System
Collector: 448 ft^2 net
Angle: 90°

Storage System
Material: 1 foot pumice block walls and 6" concrete slab

Distribution System
Direct-gain through S windows heats slab and 2-story block walls. Whenever house temperature reaches 86° F., ¼-hp fan moves heat into N rooms. At night, block walls and sun-warmed slab help warm both floors: heat comes up stairwell and through ceiling. At one time ducts were used for active air collectors on roof.

Auxiliary System
Backup: 22-kw electric resistance in ceilings
Fuel consumed; $400 at 6 cents/kwh average

Costs
House: $40,000

personally designed, installed, and lived with more collectors than anyone else in the country.

Nothing Saved

He may also be the only person who claims that his solar system "cost nothing and saves nothing." Apart from the experimental active collectors, the passive system added nothing to the cost of the house because money that would have gone into an oil furnace paid for the extra glazing, insulation, duct work, and fans. The bank required a complete duct system in case the house were sold and the new owner put in a furnace. Similarly, a furnace room was constructed to satisfy the building inspector. Norm and Jeanne use it as a root cellar to store harvest from their garden and orchard. With the various roof collectors, the house got about 60 percent of its heat from the sun. Without them, the percentage dropped to as low as 40 although the sunroom has restored it to two-thirds. For the other third the Saunderses pay electric rates, which are roughly three times as much as oil. As compared to heating with oil, they save nothing.

Pointing out that the house is a twenty-year-old design, Norm recommends that it not be built again. "We could do so much better today. I've had only one person challenge my statement that I've made more mistakes than anyone else in the field." His roof overhang cuts off the sun too early: by March only a couple of feet of floor get sun. A hill to the east shades the house until 8:30 A.M. in winter and mid-morning in spring and fall. He used metal casements instead of wood, which transmit much less heat. Worst of all, since the crawl space is uninsulated, the under-slab "storage" has been of little use. Ground water from the nearby hill flows close to the footings, continuously stealing heat. Norm urges that anyone who is burying part of a house insulate almost as well as if it were not buried.

New Designs

Building on experience gained in Experimental Manor, Norm is continuing to design new collectors. He advocates 100 percent solar heating, argu-

Norm Saunders is irritated by people who say that solar heating is not here yet. He says that Massachusetts is already getting more than two percent of its heat directly from the sun—just through windows.

ing that anything short of that will create problems for the community and individual. When people go for 50 or 60 or 80 percent, they usually put in electric heat as the lowest cost backup. The savings can be stashed in the bank and used to pay the extra cost of electricity as against oil. But suppose an entire community chooses that route, he asks. Peak load demands will get out of hand when a cloudy period hits and everybody throws the switch. From the community's standpoint, Norm argues that it will be disastrous to go for less than 100 percent solar, and from the individual's point of view it will mean enormous electric rates.

To provide total solar heating, storage has to be large and therefore bulky. The only place to put it, Norm believes, is under the roof where cheap space exists. A concrete slab by itself does not provide enough thermal mass. Storage in walls enlarges the building and thus increases property taxes: most communities tax on total exterior area. Putting storage in a basement or crawl space avoids

Looking up at the living room balcony on the south side. The wide roof overhang is so wide that it cuts off the sun too early in winter when heat is still needed.

using living area, but costs increase in that the heat must be transported down to storage. The solution proposed by Norm is overhead storage: troughs filled with water and directly heated by the sun through a translucent roof. We should point out, though, that eliminating attic space permanently removes the potential for upward expansion should your family grow.

Solar Staircase

In Norm's most interesting invention to date—the Solar Staircase™—a standard pitched roof has a heavily insulated north face but a fiberglass or glass skin on the south pitch. Beneath this skylight is a stairway that steps up the roof. The stair treads are aluminum reflectors tipped down 3° and the risers are glass. In winter radiation passes directly through the risers, or is reflected from the treads down through the risers, into translucent water troughs. The summer sun, being at a higher angle, is largely reflected out. In the first installation of the Solar Staircase—the cafeteria of Cambridge School in Weston, Massachusetts—overhead storage was omitted in favor of a massive north wall of concrete block with exterior insulation. In a later test installation, water troughs were formed by curving Sun-Lite fiberglass. Sun passing through the troughs heats the room or is reflected back to the water storage by thermostatically controlled louvers. When the room is hot enough, the louvers close; at night they open to permit heat to radiate down into the room.

In the school dining hall the added cost of the collector was $2.20 per square foot, and the first year savings were 6¢ in fuel and 32¢ in electric lighting per square foot of collector. Although the cafeteria receives continual use, the lights are never on during the daytime, whereas they're on all the time in the other rooms of the school. In summer, heat gain from the sun is much less than

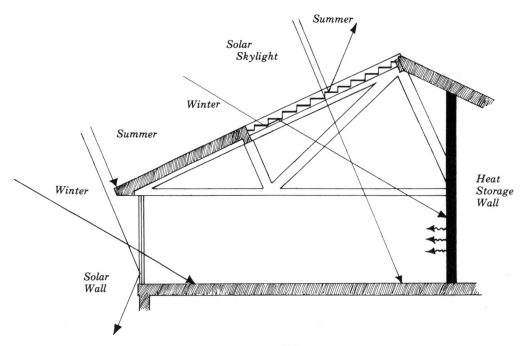

Solar Staircase™Collection System

the heat from artificial lighting would be. The light is pleasantly even and free from glare, and the Staircase creates a dramatic ceiling. Space heat savings might have been greater except that the backup system ignores the solar aspect of the building: return air registers are located high where they draw the solar heat out. "Heating engineers," Norm says, "consider energy free. I lost every battle with them."

Roof Removed

For a retrofit on the home of Roland Peterson, Norm is trying out still another invention. The roof was removed from the Petersons' one-story home so that a second floor and attic could be added. Then the roof was replaced. Since the house faces more east than south, a Staircase would not work. Instead, 420 water-filled glass bottles were placed in the attic behind a translucent east pitch and south gable. When the retrofitting is finished, fans and ductwork will move the heat down to the lower floors. At the time we visited Norm, a layer of Translucent Insulation™ was being installed beneath an outer skin or ordinary glazing. Norm asked us not to describe this invention since the patent is still pending, but he believes it will probably transmit two-thirds of the sun's rays while

cutting heat losses as compared to glass by a factor of six. If the second floor is taken as an isolated structure, it should be heated entirely by the sun and some heat contributed to the first floor, saving the owners a total of $300 a season on a $4,000 initial investment.

He hasn't done it yet, but Norm is predicting that it will be possible to build a 100 percent solar-heated home for less than the cost of a conventional house. Two things are in his favor: going with integral systems means eliminating the cost of ordinary roofing materials, and going for broke means eliminating the cost of a backup. The big

Jeanne Saunders.

Jeanne Saunders in living room.

problem is getting enough storage to insure that the solar home will cool down to an uncomfortable level only once or twice in a season, a situation which a fireplace or extra clothing could easily handle.

Temperature Swings

After nearly twenty years in their home, how do Norm and Jeanne feel about the seeming drawbacks of passive heating: large temperature swings, glare, fabric fading? Norm says he's "happy as a clam" when it gets up to 86° F. Although Jeanne becomes somewhat uncomfortable around 77° F., they both find a gradual rise during the course of a winter day more pleasant and stimulating than a constant temperature. Even the low 90s are acceptable with ventilation and appropriate clothing. Similarly, although they normally keep their thermostat at 68°, they have experimented and found the high 50s can be satisfactory.

Jeanne is outspoken about glare: "People ask how we stand this much light. If the human animal has been able to adapt to igloos in the Arctic, to no housing in the middle of the Kalahari, if Americans can adapt to brick houses with windows right next to their neighbor's, don't tell me

Looking out the living room sliding glass doors.

61

they can't adapt to this much light. It doesn't bother me at all. It's probably been good for my eyes."

But this from Norm: "Occasionally MIT students take this house as an example of glare problems, even build models and take measurements. They go away shaking their heads." Norm dreams now of a solar home that takes care of glare by using a translucent ceiling and a little less south glass—and a 60-foot-long swimming pool so he can fulfill his dream of swimming laps without frequent turns.

As for deterioration of materials, the living room cork floor is so sun-bleached that matching a tile would be impossible if it needed replacing. On the other hand, after eighteen years of wear and tear how many tiles *could* be matched anyway? An upholstered chair that has sat in the sun for four years is shot, yet its companion with a different fabric is unaffected. Thus proper choice of materials and colors can short-circuit the fading problem.

Linen drapes survive well. After eighteen years the Saunderses are only on their second set. Admittedly, the curtains are seldom pulled these days. The first eight years the family closed them religiously every night, but then Norm did some figuring and discovered that with uninsulated drapes they were saving a mere 10 cents a pull. Norm and Jeanne got out of the habit, but it stuck with the children. When they return home for Christmas, the draperies get pulled again. Laurence's conservation habits were instilled so early that he now runs his own house on solar principles: the first second-generation sun dweller.

Something Old, Something New

Who loves a garden loves a greenhouse too.

William Cowper

The neighbors may be thrifty Pennsylvania farmers, but Christopher and Melissa Fried could teach 'em a thing or two about saving money. The previous owner of the Frieds' chicken farm heated the house with 12 tons of coal a year. The Frieds use the sun and 2½ cords of wood. In 1973 the former owner spent $60 each month for electricity in the house and chicken brooder, yet the Frieds get by with $25 worth, after five years of rate increases. On top of all that, they grow most of their own food.

Pennsylvania farmhouse retrofitted with a solar greenhouse. The upper two-thirds of the greenhouse contains an air collector. Solar-heated air is ducted to a rock bin in the basement; the rocks usually heat up to 75° F.

In 1974 Christopher quit his mechanical engineering job on Long Island, taking Melissa and their baby daughter to a quiet Pennsylvania valley near Bloomsburg in search of a saner way of life. Raising 6,000 chickens a year and canning their food, they came to prosper on an income only slightly above poverty level. After retrofitting the farmhouse, Christopher found himself spending his time as an energy consultant, so the Frieds are giving up the farm to build a home nearby that will be specifically designed to demonstrate the possibilities of solar heating.

During the winter of 1974–75 their first step was insulate the attic. Next they bought 8 tons of coal for the hand-fired furnace in the basement, then discovered it was easier to operate with wood. After gathering 8 cords of deadwood from a nearby mountainside, they sold most of the coal. The house stayed reasonably warm, partly because they closed off a third of its 1,700 square feet. But the ancient furnace was so inefficient that halfway through the winter they replaced it with a wood stove upstairs in the kitchen. This heated the house almost as well with a quarter the wood. Even so, by the end of the winter they'd gone through their entire stockpile.

Christopher decided he should look into solar heating. He began accumulating books and writing for literature from manufacturers. He soon resolved to build a new wood stove and some sort of attached greenhouse. With two stories available, a simple lean-to would provide a good-sized room, even with the steep pitch called for by the latitude-plus-15° rule of thumb. He took several months to do the planning because, being an engineer, he wanted to be sure he had a flexible design, one that would give as many options as possible.

Building a structure of rafters on a foundation of railroad ties took about a week. The rafters were then covered with two layers of Sun-Lite fiberglass. An earth floor that serves as a vegetable plot is warmed by sunlight entering the lower third of the glazing. Light hitting the upper section heats the metal plate for an air collector. The lean-to greenhouse covers a large portion of the south face of the house, including the cellar door. At each end a salvaged window provides summer ventilation and access for bringing in garden tools.

The coldest temperature recorded in the greenhouse was 30° F., partly because the two layers of Sun-Lite, the absorber, and two layers of Masonite backing trap air that serves as insulation, and also because heat is retained by the earth bed and three 55-gallon drums filled with water. The Frieds grow vegetables that don't mind an occasional frost: lettuce, spinach, endive, parsley, onions—but their tomato plants conk out in January. At the time we were there, the greenhouse was under attack from aphids. Melissa had brought in ladybugs to do battle.

The absorber for the air collector consists of ninety-six used printing-press plates bought from the local newspaper for 2½¢ a square foot. Christo-

---

TECHNICAL DATA

Owner-designer-builders: Christopher and Melissa Fried, Pennsylvania

General Features
Latitude: 41° N
Degree-days: 6,000
Insolation: 115
Heated area: 1,100 ft²
Year of completion: 1975 (retrofit)
Insulation: Walls: 4" urea-formaldehyde
         Roof: 9" mineral wool
         Foundation: 3½" fiberglass
         Shutters: 1" Styrofoam on upstairs windows
Orientation: Almost due S
Solar system: Hybrid—active air and passive indirect-gain (greenhouse containing air collector)

Collection System
Passive collector: 125 ft² (lower 1/3 of greenhouse)
Active collector: 250 ft² (upper 2/3 of greenhouse)
    Angle: 56°
    Cover: Double layer 0.025" Kalwall Sun-Lite®
    Absorber: .009" aluminum, nonselective coating

Storage System
Container: 12' × 12' × 6½' wooden bin lined with plastic

Material: 20 tons of ¾"-2" diameter rocks
Location: SW corner of basement
Insulation: 6" fiberglass (top, sides), 2" Styrofoam (bottom)

Distribution System
Triggered by differential thermostat, ¼-hp blower circulates air down space between two layers of Masonite. Air then flows upward along back of absorber between fins to a manifold at the peak and through wooden ducts to rocks. Storage usually gets up to 75° F. Heat distributed by 1/6-hp blower through ducts to rooms.

Auxiliary System
Backup: 50,000-Btu wood stove
Fuel consumed: 2½ cords

Domestic Hot Water
Controlled by differential thermostat, 1/40-hp pump sends water to 40' of finned copper tubing suspended by wires in hot air manifold. Boosted by electric water heater.

Costs
House: $32,000
Solar: $1,500 (materials) plus approx. 3 weeks of owners' labor

The interior of the greenhouse, showing the translucent lower part. Sunlight strikes the earth and 55-gallon drums, which store heat to keep vegetables from freezing at night. The upper section contains recycled printing press plates fabricated into an absorber for the air collector.

pher cleaned off the ink, bent in protruding fins to create air turbulence, and etched the plates with a mild acid so paint would adhere. The absorber took two days to fabricate but cost a total of only $20. For the flexibility craved by Christopher's engineering soul, the collector is designed so it can be easily slipped out and replaced with standard units if it ever becomes desirable to do so.

Too Much Heat

The first year after building the lean-to, the Frieds operated the system without the air collector. By 11 A.M. on sunny winter days the greenhouse would become a hothouse as the temperature reached a scorching 95° F. Christopher or Melissa would then open the kitchen and living room win-

dows, both within the lean-to, and a blower would circulate hot air until late afternoon. Most of the time Christopher found himself driven out of the house. Melissa, who loves heat, had no complaints.

That year Christopher also built the new kitchen heater which he describes as a "glorified drum stove." It has a thermostatically controlled fan for blowing hot air from the kitchen into the living room. Heat reaches the bedrooms through old-fashioned ceiling registers and by drifting up the staircase. With solar heat from the greenhouse and the improved wood stove, the Frieds burned only 4 cords that winter.

In 1976-77 the main improvement was the addition of Styrofoam panels on the upstairs windows, panels which are left in place all winter. As Christopher explains, by the time they go into the bedroom it's dark outside so there's nothing to see anyway. One window is left unshuttered "so we know when to wake up in the morning." That fall the Frieds had planned to have the walls insulated; holes were drilled in the siding, but the workmen failed to show until the following May. Despite a well-ventilated house exterior, their

Christopher and the drum stove that he built and installed in the kitchen.

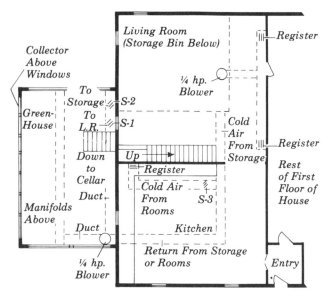

Floor plan and heating system.

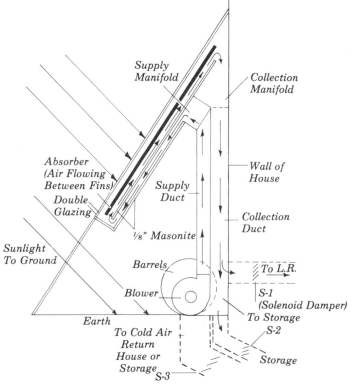

Greenhouse cross-section, showing air collection system.

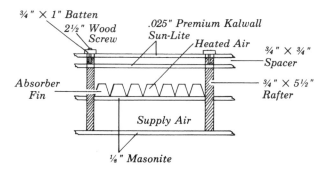

Collector Cross-Section

wood use went down a little because of the shutters.

The solar storage had also been built in the fall, but ducts and controls were not installed until late in the winter. So the fourth year, 1977–78, was the first full winter of active collection. That year they burned a mere 2½ cords. Christopher gives no credit to the wall insulation for the savings, because it shrank so badly it does little to stop infiltration. Even so the record of the active system seems slightly less impressive than that of the passive. Christopher points out, however, that once the air collector was hooked up, it became much easier to maintain a uniform house temperature.

Hard Work

The hardest part of building the system was filling the storage bin with stones. In a new house this is normally an easy job, but it can be a challenge in a retrofit. The driver backed the truck up to a basement window next to the bin, and a chute was set up. The first nine tons rolled down relatively smoothly. Inside, though, a mountain began building up beneath the window. Christopher climbed in to shove rocks by hand to the far end.

With the second load the bin was too full for the chute to rest at an angle steep enough for the rocks to slide. This time Christopher spent four hours lying on his side and stomach moving 11 tons completely by hand—and, as his hands became red and raw, by foot. He tried a chain-driven conveyor belt from a corn picker before reverting to hand labor. The conveyor worked with larger rocks, but small ones caught beneath the metal links and jammed the sprocket.

Building wooden ducts took two weeks of on-and-off labor, although Christopher estimates that a contractor could do it in a day. The ducts are plywood and Masonite and some are caulked but not airtight. Christopher prefers wood to metal because it's easier to work with, a better insulator, and a renewable resource. Where ducts are exposed to view, as in the greenhouse, wood is certainly more attractive.

For $150 Christopher bought components for a homemade logic circuit consisting of four relays. Wired into the differential and room thermostats, the controls activate two blowers and three solenoid-operated dampers. The system's operating modes are simple: solar heat is collected and sent to storage or directly to the rooms, or the stored heat is distributed. The living room window enclosed by the lean-to is partly covered by the collection duct, which contains a couple of dampers that direct solar-heated air to the living room or on to storage. The kitchen has a return duct with a damper that steers the cooled air to the collector from the rooms or from storage.

A Few Problems

Minor problems have developed. Two differential thermostats malfunctioned but were under warranty. One of the system's dampers caused us to jump as it made a startlingly loud thunk when it opened. The solenoid could be replaced with a motorized damper to reduce noise—but at four times the cost. In one of the blowers the relays somehow became magnetized and stuck. They work for the storage mode but not for the nighttime extraction cycle unless Christopher goes down and gives them a tap.

Since the extraction blower turns on only when the storage is 70° and above, distribution is frequently by natural convection and radiant heat coming up through the floor. This seems to work because this year the Frieds built only three fires from the beginning of March until the end of April. Sunny and cloudy days were evenly spaced, allowing the storage to keep up with heat demand. Christopher intuitively feels that they're getting a fourth of their heat from the sun. Because they gather their own wood, they save the full $600 a year that the previous owner paid for his 12 tons of coal.

As the chicken business is included in their electric bill, it's hard for the Frieds to tell how much they save by preheating hot water. They conserve a great deal of electricity by zone heating the chicken house. The former owner kept the entire building at 95°, the temperature required in the brooding pen. On the other hand, the Frieds use

Christopher on his solar demonstration van.

electricity for cooking whereas their predecessor had a gas stove.

Well water for the house comes in at 50° F. and is raised to about 90° in tubing located in the collection manifold of the air system. Because temperatures in the greenhouse rarely go below freezing, neither a drain-down system nor antifreeze is used. On those two or three occasions each winter when the manifold drops below 35°, the thermostat switches on the pump to circulate water.

Solar Van, Too

Christopher, subsidized by a grant from the Pennsylvania Department of Community Affairs, drives a solar exhibit van to county fairs and energy conferences around the state. The van is a miracle of compactness: it has an air collector for space heating and a domestic hot water collector mounted on one side, a demonstration of types of insulation on the other, and a wind charger on the roof to provide electricity to run the blower and pump for the two collectors. Last summer Christopher logged close to 10,000 miles going to fifty fairs. Between driving the van, doing college courses and programs for public television on energy conservation, and serving as state coordinator for Sun Day, he spends the majority of his time in educational activities.

At fairs he's found that few people understand what solar energy can do, and at what a low cost. Most have only heard of expensive active systems, and so are waiting for a breakthrough in technology—for mass production and a cost reduction. "I don't think either of those is going to come," Christopher says. "What we're seeing now is better understanding, better matching of components, refining of designs in hybrid systems, and establishment of more accurate priorities." In particular, he feels that conservation has long been an underrated priority.

In their new home the Frieds will again use a hybrid system: passive with an air collector. The house will be made of concrete block and partially underground. They're using this type of system in part because many people at the fairs find it hard to believe that passive works as well as it does. Christopher sees this as a challenge in public education, and his new home as a demonstration of the workability of passive heating. But it will take more than just building a house. After three years of looking at the Frieds' collector, a neighbor finally constructed one for himself. Most of the farmers, however, never even think to ask what the Frieds save on their heat bill. Smiling ruefully, Christopher told us their typical question is: "Does it work in the wintertime?"

Solar-Mud Restoration

Mud, mud, glorious mud!
Song

Lostetter house: View from the southwest. Part of the original turn-of-the-century structure was torn down; the rest was restored and a greenhouse was added. The ladder provides access to shutters which have two types of insulative backing, to be covered with siding.

When Al and Kathie Lostetter decided to leave Michigan for northern New Mexico, they had no intention of building a mud house. Then they began making house-hunting trips. "A real estate agent tried to sell us this ruin of an adobe house," Al reminisces. The agent told the Lostetters, "Oh you can just live in one room and redo the rest." Al's first reaction: "I know what he thinks of me—he thinks I'm a blooming fool." But the more he saw what people in New Mexico are doing with adobe, the more it seemed possible.

Al, an art professor, and Kathie, a film maker, weren't thinking at first about heating, solar or otherwise. Then they acquired several exotic birds —macaws, cockatoos—and a baby, Antony, and became concerned about maintaining even temperatures. After reading many books they settled on a design employing direct gain but still felt somewhat unsure of their choice.

Eventually they visited the Sun Dwelling Demonstration Center at the nearby Ghost Ranch in Abiquiu, N.M. The center, part of a convention complex run by the Presbyterian Church, consists of four dormitories differing only in their types of solar-heating facilities. One has a Trombe wall, another a greenhouse, while the simplest uses direct-gain through large windows on the south. The fourth or control module, has no solar heat, but incorporates energy-conserving design features. All of them work and all were constructed of adobe with pumice insulation by members of the local Native American and Hispanic communities. Seeing solar heating in use at Ghost Ranch reassured the Lostetters that they were on the right track and would be able to keep their house above the 60° required by their avian family.

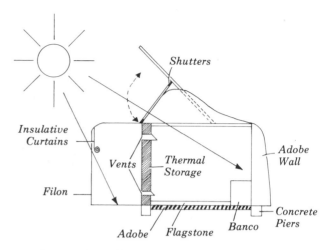

Cross section of house, showing thermal storage.

Low-Cost Housing

The Lostetter home, like the Ghost Ranch dormitories, is a model of low-cost housing. Counting land and materials, Kathie and Al have $7,000 or $8,000 invested so far. Adobe bricks cost 20¢ to 25¢ apiece new, but the Lostetters salvaged many of theirs and received some in trade. Mostly they made their own, without the aid of a mixer, in forms that hold four at a time. On a good day they could make more than 100.

They began by putting in a septic system and getting the feel of the place. Then they started to

work, fixing up the turn-of-the-century "ruin" by tearing off the old roof and removing the crumbling plaster. A concrete foundation was poured beneath the walls, and weak spots were reinforced with concrete. When adobe walls go, they fall outward, so a concrete cap had to be cast on top of the last round of bricks to tie the structure together. The aged but sound "vigas" or rafters were wrapped in plumber's tape and cast in the cap before the roof was added.

While stabilizing the structure, Al and Kathie

TECHNICAL DATA

**Owner-designer-builders: Kathie and Alvern Lostetter,
    New Mexico**

General Features
Latitude: 36° N
Degree-days: 6,000
Insolation: 250
Heated area: 600 ft^2
Year of completion: 1978 (solar restoration)
Insulation: Walls: 2" polystyrene (exterior)
        Roof: 1½" polyurethane and 1" styrene on flat
            portion; 3" fiberglass and 1" styrene on
            slanted portion
        Floor: old adobe floor and 5" mud and flagstones
        Shutters: 1" polystyrene in wood frames, man-
            ually operated (clerestory)
        Thermal drapes: Roll-up polyethylene bubble
            greenhouse insulation
            (greenhouse)
Orientation: 15° W of S
Solar system: Passive direct-gain and indirect-gain

Collection System
Collectors: 32'×6'×9½' porch single-glazed with Filon®;
    4'×32' windows on roof at 60° angle

Storage System
Adobe walls: 192 ft^3 rock-bin under greenhouse floor;
    "banco" with rock-filled drums along N wall of living room

Distribution System
Direct solar gain through sun porch heats south adobe wall
and rock storage, which radiate heat at night; direct gain from
clerestory warms other living areas. Vent system in S wall
promotes natural convection from greenhouse.

Auxiliary System
Backup: Wood stove and adobe fireplace
Fuel consumed: Less than 1 cord, 1977-78

Costs
House: $4,000
Solar: $225

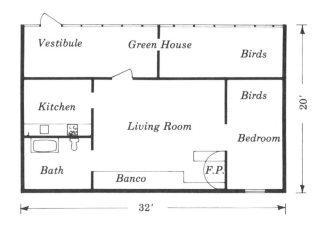

LOSTETTER FLOOR PLAN

Vestibule · Green House · Birds · Kitchen · Living Room · Birds · Bedroom · Bath · Banco · F.P. · 32' · 20'

lived in a tipi on a portion of their land near the Chama River. They liked the feel of the tipi so much that Al wanted to incorporate some aspects of its design into the adobe house. "For elegance I think the tipi ranks with cathedrals and early skyscrapers," he says. "It allows you to be out of the elements without altering your sense of exterior space, so that the outside retains its magnetism and presence."

It is in the roof that the new house manifests the tipi inspiration. The north section is flat but the southern half is a gable fronted by a clerestory running the length of the building. As in a tipi, the "window" is high above and sun comes directly through as from a smoke hole. The translucence of Filon lets the ghosts of outside images survive like the shadows on a tipi. At times a shaft of light falls on a star pattern made of diamond-shaped stones

Patting the adobe mud mixture into the forms.

in the flagstone floor, and the rectangular room suddenly seems rounder.

"We really got into mud," says Al. "It's great to work with. Nowadays people take an old box like this house was, put a frame and pitch roof on it, and cement stucco the outside—they miss all the fun." The Lostetters were convinced that the old methods were better. What if you have to do a little mud repair every year? Some of the houses in the area have been standing for a century. During the Lostetters' second spring in the house a heavy rain washed out a small gully in the north wall, but it was easily repaired with a bit of mud.

The adobe walls were insulated with an inch of styrene board first tacked on with roofing "dollars" and then secured with barn pole nails through lath. Chicken wire was stapled to the lath and the whole thing splashed with concrete adhesive. Lastly, a mixture of mud, sand, straw and wheat

Adobe mud mixture is poured into forms holding four bricks each and then allowed to dry for 3 or 4 days. Flagstones were used for flooring.

paste was troweled on in the manner traditional to the Chama Valley.

Adobe Walls Store Heat

The Lostetters believe the mud walls are helping keep them warm. The adobes, with their high thermal inertia, serve as the major heat storage. Sunlight shines directly through the greenhouse and clerestory onto the walls. In addition, half the greenhouse floor has a rock storage bin under it, while along the wall of the main room, a "banco" (adobe bench) plastered over drums filled with rock stores more heat. The banco is only one of the many adobe built-ins which the Lostetters use to conserve space in their small home.

The greenhouse, added after the main structure was basically complete, has a sunken floor. Al and Kathie felt that a single glazing plus curtains would be better than double glazing with its increased reflective losses. They use two sets of drapes: one layer of the insulating bubble plastic used for packaging and a second set of dark blue polyester shades. Last winter the porch housed bedding plants, but soon half of it will serve as a home for the jungle birds, who greet visitors with piercing cries.

The porch roof is insulated with fiberglass plus Celotex board, which provides support needed in front of the clerestory for the manipulation of shutters. With two inches of polystyrene and 1×6 framing, the shutters are heavy. But before they were installed, the sunlight was so intense inside that the baby "nearly fried." Now Kathie gladly climbs the ladder to the roof to adjust them. Once a shutter came crashing down in a high wind, but lock latches solved that problem.

The extended gable with its pointed supports leans out over the window openings and the shutters divert or catch the sun in much the same way that the smoke flaps on a tipi control the draft or keep out the rain or snow.

The adobe living room wall accumulates heat from the greenhouse and radiates it to the inside.

Vents, Not Fans

No fans are used in this entirely passive system. Air circulates between the greenhouse and the living area through vents at the top and bottom of the south adobe wall. Hot air flows from the greenhouse into the living area through the top vent, and as it cools it descends and is drawn back into the greenhouse. At night the vents are closed manually so that heat is not lost. The only auxiliary heating comes from a wood stove in the bedroom and an adobe fireplace which is not particularly energy-efficient. Last winter the interior stayed at 65° F. during the day; the greenhouse, at 90° F. On really cold nights Kathie and Al stoked the stove, but in March Al noticed that the woodpile had hardly gone down at all, so they started using the fireplace in the evenings for enjoyment. Half a cord would have carried them through.

Although they were not neophytes in the realm of construction—Al has experience at carpentry and plumbing—the Lostetters had to research adobe construction methods and solar heating, gleaning vital information from local builders. Their favorite printed sources are *Solar Greenhouse* by Bill Yanda and Rick Fisher and the *Adobe News,* published in Los Lunas, NM. Some of the information they chose to ignore: most of their sources said not to waste time trying to restore a turn-of-the-century adobe house, but the Lostetters are thankful for the savings in time and money represented by the old walls.

The interior walls were finished with a mud-wheat paste mixture in the traditional fashion, except for areas such as the kitchen and bathroom where gypsum was preferred. Different types of earth produced variously colored mud to lend variety to the rooms. Originally the walls had enamel painted over the adobe. An enamel cross on the

Kathie Lostetter with Antony. Amateur archeologists, the Lostetters are researching the use of bird feathers by Pueblo Indians in pre-Columbian times.

south wall of the living room was left for a bit of authenticity.

Ruins Nearby

Kathie and Al respect their home as part of an archaeological site. There are unexcavated Spanish-colonial ruins all around the house. Less than 50 feet away is the base of an old "torreon," or fort-tower, which they hope to restore for a studio. The first settlers had neither glass nor fiberglass and therefore were restricted in their ability to utilize the sun. In future restorations, as in their present home, Al wants to incorporate the best of both traditional and modern: "If we rebuild the torreon we will do it traditionally, but with a solar skylight so that it will look authentic from the outside but have hidden thermal efficiency."

Aware that they may have made mistakes, the Lostetters are basically satisfied with their handiwork. They toyed at first with the idea of building underground but are now happy that they decided to follow traditional design. The smallness of their house suits them because they prefer to be outdoors most of the time. Improvements they are considering include coloring the adobe wall of the greenhouse dark brown for better absorption and adding a removable awning in front to shade the birds in summer.

The floors are a nontraditional part of the house. "Supposedly prehistoric peoples used mud mixed with animal blood for the floors," Al says. Local people coat floors and walls with linseed oil. The Lostetters preferred to use flagstones gathered from the surrounding countryside. They left the old adobe floor but added 5 inches of mud on top before laying the stone. The stone floors are less dusty than adobe but present a problem for baby Tony, who is just learning to walk, so rugs are needed to provide a softer landing for his crashes.

The Rebate

Al is amazed that New Mexico allowed them a tax rebate for their homegrown solar-mud dwelling. In 1978 when the Lostetters retrofitted their home, the federal government did not give them a tax break, but the new energy law provides that tax credits are retroactive to April 20, 1977, so they will benefit. In their state tax returns, the Lostetters found that expenses for active systems are easier to deduct than those for passive, since the tax people count only materials that are clearly solar-related. Kathie and Al received an $85 credit for the wood and the Filon used in the sun porch, and the insulation for shutters on the clerestory as well as a good portion of the fenestration.

Al and Kathie make no claim to be innovators in the field of solar energy. They consider the slanted clerestory with shutters to be a new idea. Many of their neighbors are building their own homes or fixing up old ones, and some are using interesting innovations. The Ghost Ranch outreach program, they believe, has encouraged people native to New Mexico to adopt solar techniques. For example, in a little town down the road Julian Fred Vigil is building a passive solar home from plans provided by the Sun Dwelling Demonstration Center. His is one of three designed by architect Mark Chalom through a grant from the New Mexico Humanities Council to encourage low-cost housing using indigenous materials. This kind of support is what is needed to make sun dwellers of us all.

Like Father, Like Son

I have been over into the future, and it works.

Lincoln Steffens

When Tom Zaugg and his son John constructed a pair of solar homes in a residential area in north-central Ohio, they built an infant harbinger of future solar suburbs. Where today there are two solar homes side-by-side, tomorrow there may be two hundred.

Like the golf course adjoining it, the Zauggs' street is flat with few trees. The area, being one of the highest points in Ohio, is windy. Last winter the thermometer plummeted to minus 24° F., and sunny days were rare in February. With this kind of weather many of the locals found $200/month electric bills in their mail. As theirs run half that, the Zauggs—father and son—are happy with their solar collectors.

The Zaugg family is top-heavy with architects.

After getting his degree, John joined his father's architectural firm. His mother, Bernadine, is also an architect. Pat McGrew, like John's four siblings, is an exception to the family predilection: she's in women's retail.

Although they are architects, the Zauggs built typical suburban dwellings because they wanted to determine whether solar heating is economically feasible for conventionally designed houses. They were also seeing whether collectors can be integrated with the architecture of homes that have their southern exposures to the street. Tom Zaugg, in a voice that sounds remarkably like Jack Benny's, says, "It's rather easy to do a building that has the solar collector toward the back of the property, but esthetically more difficult on the main approach." Comparing the performance of a wet and a dry system was a third objective.

John's A-Frame

John's home, one of the best-designed A-frames we've seen, is finished with redwood siding. A carport leads to a family room on the north side. If you can make it past John and Pat's four-month-old puppy, you go from there into a living room with a wood stove and cathedral ceiling. High up in the roof peak, an old-fashioned ceiling fan helps circulate hot air in the winter. The kitchen and dining area are tucked beneath a second-floor bedroom and balcony overlooking the main level. Downstairs, a bath and another bedroom adjoin the kit-

---

TECHNICAL DATA

Owner-designers: John Zaugg and Pat McGrew, Ohio (A-frame)

General Features

Latitude: 41° N
Degree-days: 6,400
Insolation: 120
Heated area: 1,500 ft^2
Year of completion: 1976
Insulation: Walls: 3½" fiberglass
      Roof: 6" fiberglass
      Foundation: 2" Styrofoam
Orientation: S
Solar system: Active air

Collection System

Collector: 600 ft^2 (16 panels)
  Angle: 60°
  Cover: Double layer Kalwall Sun-Lite®
  Absorber: 24-gauge galvanized steel, nonselective coating

Storage System

Container: 38'×26'×4' concrete block bin
Material: 100 tons rocks 1"-2" diameter
Location: Crawl space
Insulation: 2" Styrofoam on all sides

Distribution System

½-hp blower circulates air through collector to storage, where 1/3-hp blower sends hot air through ducts to house.

Auxiliary System

Backup: 17,000- Btu electric resistance and wood stove
Fuel consumed: $200 at 2½ cents/kwh average and 1½ cords at $50/cord

Costs

House: $45,000
Solar: $3,000

Owner-designers: Thomas and Bernadine Zaugg, Ohio (Ranch-style)

General Features

Latitude: 41° N
Degree-days: 6,400
Insolation: 120
Heated area: 1,500 ft^2
Year of completion: 1976
Insulation: Walls: 3½" fiberglass
      Roof: 6" fiberglass
      Foundation: 2" Styrofoam
Orientation: S
Solar system: Active liquid drain-down

Collection System

Collector: 768 ft^2 (36 panels)
  Angle: 60°
  Cover: Single layer Kalwall Sun-Lite®
  Absorber: 28-gauge stainless steel, nonselective coating
  Pump: Two ½-hp pumps

Storage System

Container: Two concrete block tanks waterproofed with Thoroseal

Material: 8,000 gal. water
Location: Beneath garage floor
Insulation: 2" Styrofoam on all sides

Distribution System

Pump circulates solar-heated water to heat exchanger in heat pump. Blower in heat pump sends heated air through ducts to house.

Auxiliary System

Backup: 35,000-Btu electric resistance and heat pump
Fuel consumed: $411 at 2½ ¢/kwh average (resistance heaters and heat pump)

Domestic Hot Water

Filtered rainwater feeds solar storage. Electric water heater draws directly on solar storage for preheated water.

Costs

House: $70,000
Solar: $7,000

The A-frame and its huge collector. Courtesy of the Zauggs.

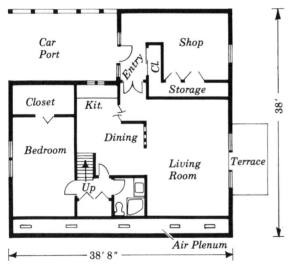

FIRST FLOOR PLAN

Car Port

Closet

Bedroom

Kit.

Entry

Cl.

Shop

Storage

Dining

Living Room

Terrace

Up

38'

38' 8"

Air Plenum

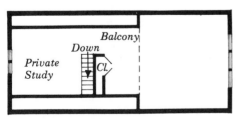

SECOND FLOOR PLAN

Balcony

Down

Private Study

Cl.

chen, adding up to a compact floor plan perfect for two people and a puppy.

The air collector is composed of 20-foot-long panels inset between the rafters. Like the designer of Sarah Sanford's home (see p. 42), John felt that the gain from fins for producing air turbulence is too small to justify the extra expense and complexity. He thought the system "should be simple enough for contractors to put it together without much supervision." Air from the rocks in the crawl space is fed into a plywood supply plenum at the bottom of the collector and flows up the face of the absorber to a sheet metal duct contained in an attic that's just big enough to enter.

The Ranch Style

Next door, Tom and Bernadine's home resembles a ranch-style house with the addition of a partial second floor containing a guest room and work space. The first story is filled with the handsome products of Tom and Bernadine's hobbies: paintings, handcrafted furniture, stained glass, and

Tom and Bernadine Zaugg in front of their trickle collector.

jewelry. Bernadine, who built a highly finished desk in high school shop long before women were taking up carpentry, designed and constructed an eye-catching barnboard banister for the stairway.

Their trickle collector, which replaces roof shingles, contains stainless-steel ridges spaced far apart to spread the water over most of the surface. A large garage on one end of the house contains the solar equipment: storage, heat pump, and controls. The stored water is separated into two tanks sunk

into opposite sides of the garage floor. Tom strongly recommends double storage because of its flexibility. Water can be taken or returned from either tank, depending on how the valves are set. Since a heat pump works most efficiently with warmer water and a collector most efficiently with colder water, a hot and a cold tank were established. Also, the heat pump can cool the house by circulating water from the cooler tank through the heat exchanger into the other tank, in the process cool-

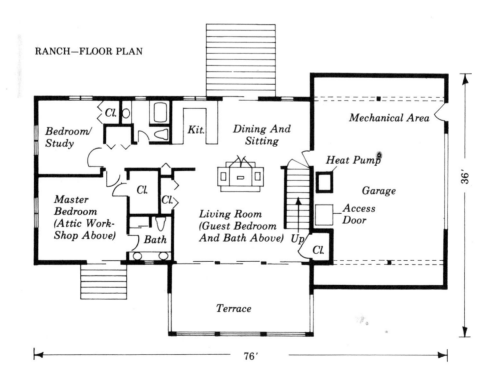

RANCH—FLOOR PLAN

ing the air going to the house. The Zauggs find the system works well, even in a humid climate like that of central Ohio.

A Comparison

Doing performance comparisons has generated a little friendly competition between generations. John conservatively estimates that the air collector is providing about 25 percent of his heat and claims that his parents' collector is doing about the same. But Tom says that, when preheating domestic hot water is taken into account, they're getting closer to 55 percent. For an air collector they're doing for a client, the Zauggs predict an improved performance of 40 percent because they're correcting mistakes made in the A-frame.

They also calculate that the ranch has a 30 per-

Balcony in the A-frame overlooking the living room. Note ceiling fan to push down heat.

cent greater heat load than the A-frame. Not only does it have more windows, it has less insulation because in the A-frame the heavily insulated roof replaces two walls. Also, the elder Zauggs set their thermostat at 71° F. day and night in order to provide a steady temperature for instrumentation installed by the local power company. John and Pat keep theirs at 70° during the day and 60° at night. In comparing performance other factors that have to be weighed are the larger surface area of the water collector versus the extra glazing on the air system.

Without sophisticated instrumentation on both houses, any judgment of relative performance has to be inconclusive. Only one thing is clear: the two are close enough in performance that the extra cost of the water system is hard to justify. Of course, a trickle collector can be done less expensively by using corrugated aluminum rather than stainless steel—selected for durability—by replac-

Sun porch in ranchhouse. This house receives some passive solar heat, whereas the A-frame has no windows to the south. It does get some early morning and afternoon sun.

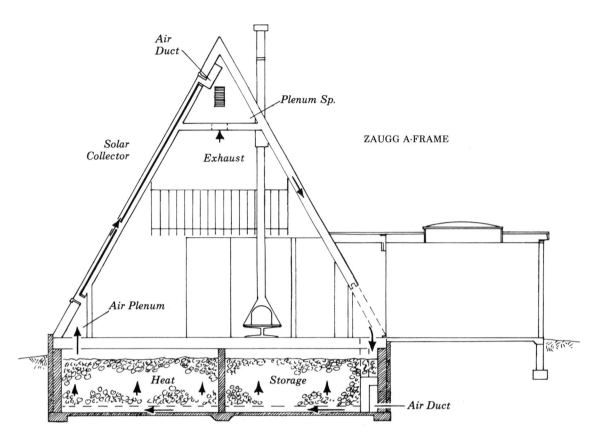

Air
Duct

Plenum Sp.

ZAUGG A-FRAME

Solar
Collector

Exhaust

Air Plenum

Heat   Storage

Air Duct

Top: *Schematic of A-frame collection system and rock storage. Below: schematic of ranch collection system. The Zauggs feel comparison between the two systems has been valuable.*

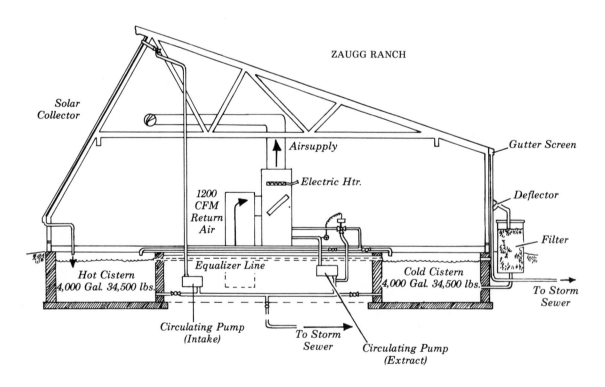

ZAUGG RANCH

Solar
Collector

Airsupply

Electric Htr.

1200
CFM
Return
Air

Gutter Screen

Deflector

Filter

Hot Cistern
4,000 Gal. 34,500 lbs.

Equalizer Line

Cold Cistern
4,000 Gal. 34,500 lbs.

To Storm
Sewer

Circulating Pump
(Intake)

To Storm
Sewer

Circulating Pump
(Extract)

ing the heat pump with water-and-rock storage and by reducing the amount of storage. But compared to the Zauggs' $7,000 version, the air collector has a decided advantage in initial cost, although the $3,000 spent on it would easily cover a less sophisticated trickle collector.

Preheating domestic hot water and providing air conditioning increase the savings produced by the water system. At the same time, the water collector should prove to have greater operating and maintenance costs. Since air systems are simpler and cheaper to build, operate, and maintain, Tom Zaugg recommends them for the general public, although for himself he's committed to a water collector because of its versatility.

Will Build Another

He and Bernadine are planning to sell their home and build another to get more space for their crafts. For the new house Tom leans toward a water collector feeding into pipes beneath the floor: radiant heating. Before building the ranch and A-frame, the family lived for many years in an essentially passive solar home that Tom built. Half the south wall was glass protected by an overhang designed to shut out summer sun. Water flowing through heating coils in concrete and brick floors helped store the heat collected, which was supplemented by a boiler supplying water at 90° F. With the entire floor as a radiator, the water temperature can be moderate, making this distribution system ideally suited to the low temperatures delivered by flat-plate collectors.

"In the 25 years that we owned that house," Tom says, "it saved us about $6,000 worth of fuel"—this

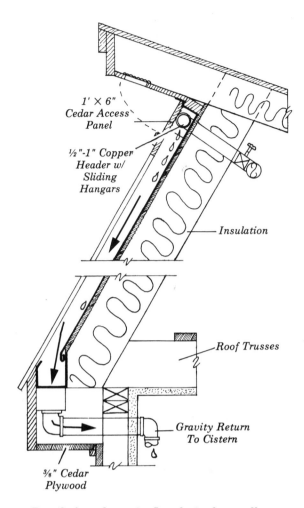

Detail of ranch gravity flow drain-down collector

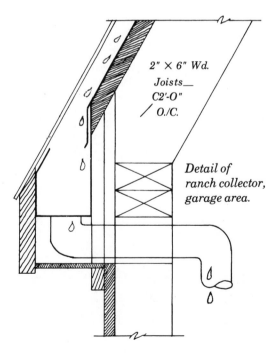

Detail of ranch collector, garage area.

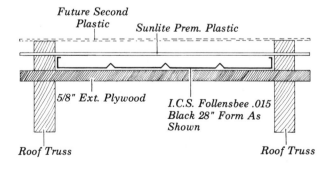

Section through ranch collector.

is in the days of ridiculously cheap fuel. Bernadine feels that the passive house was more comfortable than the ranch because radiant heat is free from the drafts produced by forced-air.

Smaller Storage

For clients, however, the firm is sticking to air collectors with some passive gain also incorporated. In the client's house under construction, the collector has only one cover but air will be circulated behind the absorber instead of in front so as to retain a dead-air space. In addition, the rock storage was decreased by half—from 100 to 50 tons. The Zauggs find that, although a huge storage allows heat to build up in August and September for use in October and November, it takes a long time to warm up—a disadvantage in February when sunny days are few and far between. It's a trade-off, but Tom and John are opting for smaller storage.

Pat McGrew, John Zaugg, and puppy.

Another major change is the use of a "reverse flow" system in the storage. Solar-heated air will go through one direction, heating the rocks at the point of entry, and air for heating the house at night will reverse the flow, coming out at the same point. This permits the hottest air to enter the rooms, whereas in the A-frame solar-heated air goes in one end of the storage, cools, and goes out the other to the house.

A reverse flow also makes it easier to heat the house directly off the collector. Although this can be accomplished in the A-frame by manually setting dampers, it lacks the finesse of the new setup in which a blower automatically pulls heated air off the collector to bring the house up to temperature before sending the air to storage. Adding complexity, however, raises costs—the exact amount is buried by corners cut elsewhere. For example, the Zauggs found that installing solar equipment in the garage is cheaper than integrating it into the structure as was done in the A-frame.

The rocks, John now believes, should not preempt the entire under-floor area: he would like to use that space for a basement with solar storage relegated to a corner or put in a garage. Rocks, incidentally, can take a long time to dry out if they start out wet. John and Pat's huge bin dried in six months, but their houseplants ate up the humidity.

Resale Value

Like his parents, John is planning to sell his home, but for a different reason. He and Pat are shifting locales. Father and son express differing opinions about the salability of solar dwellings. Although the elder Zaugg believes his collector will increase the value of his home within two or three years, "right now it would be a borderline situation." John, on the other hand, feels that his will help the sale today. One owner we interviewed pointed out that a solar collector probably won't affect resale value as much as the number of bedrooms. But a publication from HUD predicts that in a few years solar homes may be selling at a premium.

Father and son also disagree about the future of solar energy. John would like to see the country invest in orbiting power plants, which would transmit energy back to earth for heating homes. That way, he argues, all the technological problems would be concentrated in one place rather than being multiplied in thousands or millions of homes. Moreover, a solar satellite would be more efficient in that it would receive four times as much sunlight as our rooftops, due to 24-hour daylight and absence of cloud cover. John further maintains that generating solar power or any other form of energy is more efficient the larger the scale.

The assumption that bigger is better, however, is questioned by many economists. Transmitting energy from huge plants over long distances to consumers involves such enormous losses that the energy industry itself is starting to consume much of our national growth in energy production. Similar transmission losses would be likely to wipe out the satellite's headstart in sunlight.

"Solar energy," says John's father, "is most practically applied at the residential level." Even if a space station is nice to dream about, it won't be tried for years. Perhaps individual collectors are less glamorous and can solve only a part of the nation's energy problem, but they're here now and they're something the individual can do for himself. Solar suburbs are a place to start.

Twentieth-Century Cliff Dwelling

We can all see in the day, and this seeing is sacred for it represents the sight of the real world which we may see through the eye of the heart.

Black Elk

The notion of living in a cave stirs atavistic longings in many of us, deep memories of our distant ancestors. Charles and Betty Nystrom brought our

The cliff-dwelling is in a sandstone bluff a mile from the Colorado River. Its chimney was constructed of sandstone blasted out to form the cave.

primitive past into the twentieth century when they built a modern cliff dwelling on their canyon property near Loma, Colorado. They took their inspiration from the Pueblo Indians, who for centuries built stone and mud dwellings into caves in the sides of canyons, orienting them to catch the sun in winter and breezes in summer. The Nystroms were lured by the promised benefits of the constant temperature maintained inside a cave rather than by the possibility of solar heating. But the bluff that seemed appropriate for their venture had a southwestern exposure. After living in their cliff dwelling for a winter with only a fireplace for backup during the coldest month, the Nystroms realized the effectiveness of their simple passive solar-heating system, based on direct-gain.

Using a large expanse of glass, nearly 300 square feet, was a means of building a "cave" that didn't feel like a cave. In researching the project, Chuck visited many "cave houses" but found them depressing: "People would just blast a hole and then spray concrete on the walls, so you really felt as if you were in a cave." The challenge for him became to combine an energy-saving approach with modern construction methods to create a house in which the average twentieth-century American would feel comfortable.

From the inside the house looks like a conventional frame structure, except that the only windows are on the southwest wall that covers the mouth of the cave. This wall consists almost entirely of glass on all three floors. The windows and doors provide more than enough light to prevent the inhabitants from turning into cavemen.

The cliff-dwelling has a southwest exposure and is partially shaded by the bluff and the decks.

Too much sunlight would be a problem, both in terms of heat and of glare if it were not for see-through Mylar shades, which have a gold reflec-

---

TECHNICAL DATA

Owner-designers: Charles and Betty Nystrom, Colorado
Builders: Mike Davis, Scott Nystrom

General Features
Latitude: 39° N
Degree-days: 5,600
Insolation: 210
Heated area: 1,925 ft^2
Year of completion: 1977
Insulation: 6" fiberglass on SW exterior wall where not glazed; house is surrounded by 2'-3' of dead-air space inside cave
      Shades: Reflective Mylar
Orientation: 30° W of S
Solar System: Passive direct-gain

Collection System
Collector: 300 ft^2 SW glass wall, double-glazed
  Angle: 90°

Storage System
House is built into man-made cave—heat is stored in mass of house and sandstone

Distribution System
Radiation and natural convection

Auxiliary System
Backup: Westinghouse heat pump and two fireplaces
Fuel consumed: $27 at 3½¢/kwh average for January, 1977

Costs
House: $60,000
Solar: Negligible
Blasting: $9,000
Heat pump (including ductwork): $3,500

tive surface on the outside. Although they cost around $70 apiece wholesale, they are well worth the price, for without them the house would become a solar oven, what with the westerly orientation. Wide balconies, which create more open space, also provide shade during the summer. Since the windows open at the bottom for ventilation, the auxiliary heat pump is not overworked in its cooling mode.

Nor does the house require much additional heat other than that provided by the windows in winter. Chuck figures that his heat pump has a much higher capacity than necessary: "I read everything I could get my hands on about underground housing, earth-integrated housing, cave houses. But nothing told me what to do. So I overdesigned everything." Because there is a separate meter for the heat pump, the Nystroms know that in January of 1977 it cost 80¢ to 90¢ a day in electricity to maintain an interior temperature of 70° F. It was discovered that the heat pump was installed incorrectly, and it had to be replaced at the end of January. The family got along fine during its absence with help from the fireplaces, even though the Mylar shades are not very effective in preventing nighttime heat loss.

Blasted Own Cave

Betty and Chuck, a retired building contractor, came up with the idea for the cliff dwelling while floating down the nearby Colorado River in a fishing boat. A fellow angler looked up at the cliffs and remarked that it would be nice to glass in one of those caves for a fishing cabin—like a modern Indian. Becoming fascinated with the idea, the Nystroms spent three years designing the house; the construction took about six months.

Carving the cave out of the sandstone was the first task. A mining contractor was hired to blast the excavations, one for the separate garage to the northeast, and another for the house itself. The latter—30,000 cubic feet—took two months and cost $9,000 to blast. After that, everything was easy.

Concrete piers rise from the rock floor of the cave to support joists above the crawl space. On three sides the house is framed with 2×4's on a 16-inch

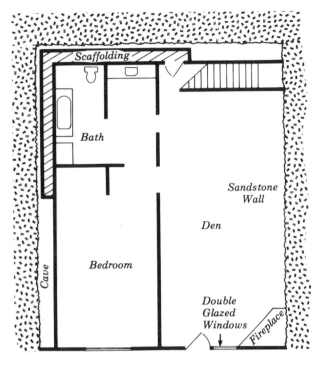

FIRST FLOOR PLAN

center and covered with plywood sheathing. No windows, no insulation. The ceiling consists of dry wall applied to 2×6 joists. If you could enter from the rear you would never know you were in a cave, for all the interior walls are finished with Sheetrock and paneling or wallpaper. One spectacular sandstone wall remains in the first-floor living room, highlighted by recessed spots and hung with greenery.

Exiting through a "back door" we were invited to walk around the exterior of the house inside the

Sandstone wall has indirect lighting.

SECOND FLOOR PLAN

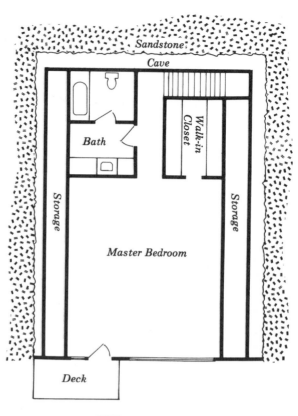

THIRD FLOOR PLAN

cave. To the observer gazing up three stories, the size of the cavern is awesome. The space around the house is used to good advantage too. Plumbers and electricians love this house, Chuck says, because they have complete access to the utilities and the ductwork. The controls for the heat pump are also located in the rear. In this way no costly floor space is given over to a mechanical room.

The Nystroms are convinced that their cliff dwelling is economically competitive with either a conventional home or one with an active solar system. The cost of blasting was largely offset by the savings derived from building inside the cave. There were no excavation or backfill costs, no footings to pour because the house sits on solid rock. The walls are without exterior finish, windows, or insulation. "Everybody else has to deal with roof trusses, shingles, and gutters. When I get to my ceiling, I'm through," Chuck chortles.

Energy Savings

And, of course, the energy savings are substantial. "The sandstone is an excellent temperature maintainer," Betty explains. "We can heat and cool the house for a quarter the normal cost for this much space." A natural cave will maintain an internal temperature of around 55° F., but their solar-heated cave stayed between 63° and 72° during the winter without the heat pump. The fireplaces were used during the evenings, but they are not designed for energy efficiency.

Heat Pump

"Now, for an efficient, functional, active solar system, you're talking about $7,000 to $10,000 on top of normal construction costs, because you have to have a backup," Chuck says. The Nystroms figure they paid nothing extra for their solar heating, while the heat pump system, including the duct work and registers, cost $3,500.

In simplest terms, a heat pump moves heat from one place to another. It takes heat from a low-temperature heat source and pumps it to a heat sink. The ordinary household refrigerator is a heat pump, in that the heat inside the compartment is dumped out into the kitchen. Likewise, in a heat pump used for space heating a liquid refrigerant absorbs heat from the outside and the resulting hot gas is pumped indoors by a compressor. The gas passes through a heat exchanger, and air blown across the hot coil circulates through ducts to the house. As the vapor cools, it becomes liquid again and returns to the outdoor coil. In the summer the process is reversed to remove heat from the house. Heat pumps use as refrigerants fluorocarbons that contribute to the destruction of the ozone layer of the atmosphere.

Steve Santoro, chief solar engineer for Acorn Structures, recommends that heat pumps be used only where air conditioning is required because the actual dollar savings don't live up to manufacturers' claims. It's true that a heat pump uses electricity two to four times more efficiently than resistance heaters, but this efficiency goes down as the outside temperature decreases. At below 20° it becomes as costly as an electric resistance system. This problem need not affect a heat pump used in conjunction with an active solar collector, since in that case it upgrades the low-temperature heat produced by the collector. But even when they save power, heat pumps don't save as much money as you'd expect.

In keeping with the inspiration for the house, Betty has employed a Southwest Indian motif in

"Moss rock" fireplace built by Scott Nystrom.

the interior decoration. A fieldstone fireplace downstairs has an air intake cut into the stone like a "sipapu," the Hopi traditional means of allowing the evil spirits to escape from the kiva. Aboriginal patterns appear in the bedspread in the master bedroom, are repeated in a wall painting, and show up again in wallpaper in the twentieth century all-electric kitchen.

Outside, a stairway carved into the sandstone provides access to the upper levels of the house and to the balconies, which are used often for barbecuing. We watched a family of chipmunks on the

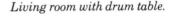

Dining room with light fixture made by Chuck Nystrom using pottery with Native American designs.

Living room with drum table.

Steps carved into the sandstone bluff lead into the cliff-dwelling.

second-floor balcony. Chuck had constructed a run for them across the front of the house from the bluff to a railing. With the reflective Mylar shades drawn, they were oblivious to our stares as they scurried up to claim their ration of nuts and seeds.

Sometimes Betty makes "sun tea" in a glass jug on the balcony. "You don't need hot water. Just take what comes out of the tap and put some mint or lemon grass in it. Come back in a couple of hours and it's brewed."

This is the last house that the Nystroms expect to build. Chuck is retiring from the construction business with a feeling of accomplishment. "This is one of those few times when you have an idea that something will work and it actually does," he says. He is particularly proud of the fact that the cave house occupies ground that is not agriculturally productive.

The type of sandstone formation appropriate for such a structure is found throughout the western United States from Montana to Mexico. Although he is not interested in participating actively in more construction, Chuck does consulting work for other potential cliff dwellers. One day there may be modern solar-heated pueblos in the Great Southwest.

Sunship

*He had been eight years upon a project for extracting sunbeams
out of cucumbers, which were to be put in vials hermetically
sealed, and let out to warm the air in raw inclement summers.*

Jonathan Swift

When Roslyn and Phil Barbash built their New York home, "Sunship," it had the first residential installation of an evacuated-tube collector in the United States. Owens-Illinois, which manufactured the tubes for the Barbash collector, had previously marketed them only for public installations such as Terraset Elementary School, Reston, Virginia, and for commercial uses such as the processing of beer and soup. Other manufacturers including General Electric and Corning Glass Works are also developing these tubes.

Using the same principle as a thermos bottle, an evacuated tube consists of a double-walled tube sealed hermetically to maintain a vacuum. A sep-

arate inner tube serves as feeder for the heat-transfer medium (water). The vacuum provides insulation by reducing heat loss from convection and conduction, and protects the selective coating by preventing deterioration from moisture and oxidation. Glass tubes are used since a vacuum would be difficult to create between flat plates.

The tubes are capable of heating water under pressure to 240° F., more than hot enough for space heating but also sufficient for operating absorption air conditioners. The latter require high temperatures in order to vaporize liquid as part of the refrigeration process. The Barbashes use their evacuated tubes for space heating alone. With their Long Island home only a few yards from the Atlantic, Roslyn and Phil prefer ocean breezes to artificial cooling.

The tubes are efficient collectors of diffuse radiation. While visiting the house during a March storm, we were startled when the collector started up despite a heavy overcast. After fifteen minutes water was returning to storage at 125° F.

The Barbashes had to pay the research price for the tubes, as well as a prototype's typically high installation costs. Commercial feasibility would depend on lowering the tag to $10 a square foot for uninstalled tubes. In the opinion of Phil Barbash, a semiretired consumer credit expert who headed an insurance corporation, the assembly line is not far off. "Owens-Illinois will have to build a factory, which would cost between 15 and 18 million dollars. The scuttlebutt from the engineers is that O-I, to avoid tying up cash in inventory while developing the market, is trying to persuade the Department of Energy to stockpile the tubes."

A technical problem has held up mass production for at least a year. As long as water can be kept circulating in the tubes, no problem exists. But under certain conditions—such as a power outage or pump failure—most of the water may drain out, allowing a tube to heat to 600° F. The few drops of moisture left become superheated steam that blows the tube apart.

Although the Barbash collector has escaped such explosions, the inventor spent a weekend there deliberately simulating the conditions necessary to blow a tube. Phil found pieces of glass 15 to 20 feet away. "My wife was concerned in case this should happen with our grandchildren around." A fail-safe mechanism was installed: when temperatures approach 220° F., a solenoid valve opens, forcing cold water from the main to flood the tubes. Since no electricity is involved, a pump or power failure would have no effect. A solution like this would not work where water pressure is less than 60 pounds, as in rural areas

---

TECHNICAL DATA

Owners: Philip and Roslyn Barbash, New York
Designer: John S. Whedbee, AIA

General Features
Latitude: 41° N
Degree-days: 5,000
Insolation: 130
Heated area: 3,000 ft^2
Year of completion: 1976
Insulation: Walls: 6" fiberglass
  Roof: 9½" fiberglass
  Foundation: 3" fiberglass
  Shutters: Insulated drapes on N windows
Orientation: S
Solar system: Active liquid

Collection System
Collector: 450 ft^2 net: 384 evacuated tubes
  Manufacturer: Owens-Illinois, Inc.
  Angle: 57½°
  Cover: Each evacuated tube consists of a glass 2"-diameter outer tube with double walls: a vacuum in the 3-mm space between the walls serves as insulation.
  Absorber: Selective coating on inner wall of the vacuum tube. A 0.6"-diameter center tube feeds water in from the manifold; it flows back through the space between this feeder tube and the vacuum tube.
  Pump: Tubes grouped into 4 modules, each served by a 2½-amp pump, plus a 1/12-hp pump for additional lift

Storage System
Container: 6' diameter spherical fiberglass tank, half-buried in sand
Material: 1,000 gal. water
Location: SW corner of basement
Insulation: 3½" fiberglass

Distribution System
A pump triggered by thermostat circulates water from a copper coil inside the storage tank to heat exchangers in the ducts. ¼-hp blowers distribute hot air to 5 heat zones in the house.

Auxiliary System
Backup: 120,000-Btu and 64,000-Btu 60-gal. propane heaters and two fireplaces with water grates, all adding heat to the solar storage through another copper coil.
Fuel consumed: 2,000 gal. propane and 1 cord wood

Domestic Hot Water
Preheated in another copper coil, ½" with extruded fins, submerged in storage tank.

Costs
Solar: $25,000
Operating costs: $15-18 a month for pumps and blowers

Evacuated tubes such as these are being used by Anheuser-Busch for preheating water in the processing of beer and by the Campbell Soup Company for processing soup. They form a dramatic section of the Barbash home.

that have well water. An alternative would be a standby battery-operated pump.

Vandalism could conceivably be another problem, but replacement of a broken collector is relatively simple because the tubes are arranged in modules. Since the tubes are strong, the Barbashes probably worry needlessly about vandalism from gulls. Watching seagulls dropping clams on nearby rocks to crack the shells, Phil and Roslyn wonder how long it will be before they decide to bomb the collector.

Little Maintenance

Maintenance is proving simple. Roslyn, a retired physician and amateur sculptor, is pleased that as yet nothing seems to affect the tubes: blowing sand, salt spray, hail. Snow melts off immediately, and rain keeps the collector clean enough to make washing unnecessary. During a first-year maintenance check, an Owens-Illinois engineer tested every tube, tightened a nut on one and that was it. O-I believes the tubes will have a life of about twenty years. They survived Hurricane Belle with

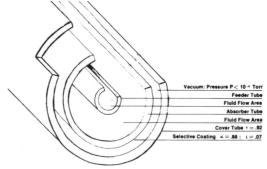

The Owens-Illinois evacuated tube.

flying colors, although the Barbashes were evacuated because high tides were expected.

Resembling a ship afloat in a grassy salt marsh, the shore home with its weathered cedar siding is flanked by clay tennis courts and a solar-heated swimming pool. During spring and fall when the heat is not needed in the house, solar-heated water is piped to the pool to extend the swim season. A blower inflates a nylon pool cover until enough headroom exists to permit entrance through a zippered door.

The lower level of the house contains a bathhouse on the east end, a carport in the center, and a studio, mechanical room, and laundry on the west side. With the carport as a virtual wind tunnel, together with the size of the building, its high ceilings and large expanse of north glass for ocean views, the house is not ideally designed for solar heating. Phil estimates that it would cost a whopping $2,000 to $2,500 a year to heat conventionally, so the current cost of $800 to $1,000 for wood and propane (includes clothes drier and kitchen stove as well as auxiliary space heating and hot water) means that the sun is providing at least half the heat load.

Oak paneling is used liberally to line the second-floor living area and third-story crow's nest, and ladders to two of the lofts reinforce the feeling of being in a ship. A sizable greenhouse off the living room supplies passive heat. Roslyn regrets that no provision was made for distributing that heat, and also wishes that the greenhouse were designed for easy shuttering.

Principles Involved

Youthful in demeanor and attitude, the Barbashes came to solar heating out of sincere environmental convictions. Roslyn describes Phil as one of those people who "saves everything, reuses everything." She has been active in the fight against air pollution since 1966. Practicing as an allergy specialist in New Jersey, she became concerned about the effect of polluted air on asthma patients, a concern that led to creating a county-wide organization made up of her patients, which eventually became the New Jersey Citizens for Clean Air. When she retired from her practice to devote herself exclusively to the battle, she was appointed by the governor of New Jersey to a state advisory council and served as chairperson of the environmental health committee of the state medical society. "Solar energy first came to our attention," she says, "as a

GROUND FLOOR PLAN

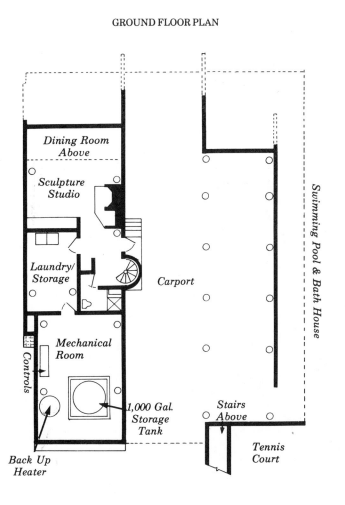

MAIN FLOOR PLAN

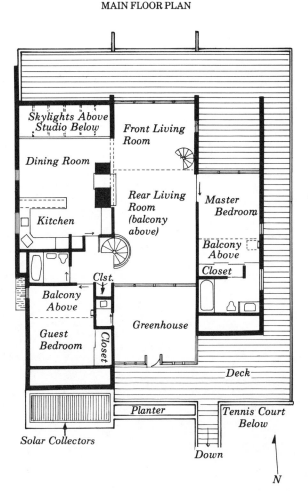

clean source of energy and secondly as a means of supplying energy without being reliant on fossil fuels, or on nuclear plants, which we had strongly opposed for a long time."

Phil adds that they were persuaded of the need for some individuals to gamble. "If you were to rely on a lot of befuddled bureaucrats in Washington, you could wait for another twenty years to get anything done." He explains that at the time they were planning to build, the Housing and Urban Development (HUD), entrusted with presenting a plan to Congress by January of 1976, was in a state of turmoil because the Energy Research and Development Administration (ERDA) had just been created. People in HUD didn't know what their jobs would be, who would have jurisdiction over residences, what the program was going to be. "So we acquired a site, proceeded to have plans drawn, and were determined that if the government wasn't going to do something that by God we were." To prove that solar energy is here, he and Roslyn would gather statistics on the house, then go down to lobby in Congress, "making ourselves such darned nuisances that they would do something to get us out of their hair."

The house plans called for a large roof area for flat-plate collectors. About this time Phil learned

The crow's nest is intended to entice three grandchildren to visit. One is at Swarthmore, another at Hampshire College, and the third at Brown. The crow's nest has a deck overlooking the tennis court.

of evacuated tubes. He phoned Owens-Illinois, a meeting was arranged, and eventually the Barbashes were assigned the necessary number of tubes out of O-I's limited production for the year 1975—no guarantees or warranties included in the price. Phil and Roslyn had planned on instrumenting the house to get data for their crusade, but Owens-Illinois volunteered to monitor it for a year. Phil has since attempted without success to

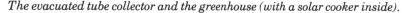

The evacuated tube collector and the greenhouse (with a solar cooker inside).

Section through Barbash home.

get information on the collector performance from Owens-Illinois.

Engineers Blamed

Along with this disappointment came disillusionment with engineers. "They don't know nearly as much as you think they do when you first listen to them. You're overawed by all their technical terms and you fail to ask the questions you should." Roslyn adds that she discovered that engineers are not infallible: "They're human and very capable of making terrible errors."

For example, common sense says that the highest array of tubes, 35 feet above ground, would put the greatest load on the pump and should receive first call on any water. But not only did the engineers put in one large pump to feed the four arrays, later replacing it with four smaller pumps that were less noisy, they also arranged the manifold so that each group of tubes was fed separately, with the highest receiving water last. Since a pressure drop occurred by the time the lower three arrays were fed, the top set of tubes suffered from starvation. Still another pump had to be added. Roslyn says the engineers "would pass the buck a lot, saying, 'This really isn't in my field; this was his— he should have known.'"

Moreover, the buck-passers wanted the storage tank half-buried in sand but otherwise uninsulated, claiming that the heat could simply dissipate through the house. What about summertime? When the mechanical room heated up like a ship's boiler room to 135° F., Phil and Roslyn added in-

sulation themselves. Incidentally, although the tank is experimental, the Barbashes were reassured about trying it when they learned that fiberglass tanks are used to store sulphuric acid. Fiberglass is immune to corrosion and cheaper than steel, while the spherical shape is a materials saver.

The engineers made other mistakes. In designing duct work running to the master bedroom, they failed to take into account the length of the run and the open port between the mechanical room and bedroom. The latter was difficult to heat until the Barbashes hooked up another propane heater. Actually a second heater had to be added anyway because the original was remarkably inefficient. Although rated at 96,000 Btu, it delivered only 64,000. When the inefficiency was pointed out, the manufacturer took a heater off the production line and had it specially baffled, so that the Btu in heat coming out would be closer to the Btu in propane going in. The loss on the new heater is only 5 percent.

Other errors range from the really inept—the plumbers installing ¾-inch pipes throughout the house, then using half-inch copper coil for the solar preheat, thus inadvertently reducing the water flow—to areas where design work is still needed. Phil believes that Owens-Illinois should manufacture a unit to connect the end of the manifold with the internal plumbing, designing it so that no possibility of freezing exists. "At present," he says, "it's been done too much on a catch-as-catch-can basis." A copper pipe wrapped with insulation is used to prevent freezing, creating a weak point in

The east view of the Barbash home shows its resemblance to a ship at sea.

the system since it means relying wholly on the care taken by whomever does the insulating.

The Barbashes wanted people to see the system and duplicate it but are now fearful that its cost and complexity may defeat their purpose. Phil takes responsibility for some of this, because he wanted to try for 100 percent solar heating by concentrating all the solar heat in the master bedroom during the three coldest months. To retreat to the bedroom and leave the rest of the house at 55° F. requires different temperature zones, so the engineers came up with no less than five blowers.

Not a Prototype

Admitting that a house this elaborate cannot be a prototype for future solar homes, the Barbashes nevertheless believe that it can serve as an example for others to learn from while building simpler systems. Phil is "thoroughly convinced of the effectiveness of the evacuated tubes."

Speaking as an expert in credit financing, he is also persuaded that solar installations are one of the soundest investments an individual can make. Consider a person who has ten years left on his mortgage. Rather than selling that house and using the profit to buy another, how much better off that individual would be to take out a home improvement loan to finance a solar installation. The loan would be repaid with fuel savings and in a few years the whole system would be paid for. From a

retirement point of view he's better off ten years hence with a house not only free and clear, but also with a source of free heat in that house when he's on a fixed income.

The banking world, Phil believes, is changing its attitude toward financing solar retrofits. Because of the bitter competition that exists in the money field among banks, savings and loan associations, and insurance companies, there is need for the small-loan customer. The default factor is lower than with large borrowers. "Back in the early 70's the banks got into real estate investment trusts and lost their shirts. The attitude has finally prevailed that the individual consumer is a better risk. Pick up any newspaper today and you'll see how anxious banks are for home improvement loans."

Don't Wait

The Barbashes clearly are accustomed to discussing the pros and cons of solar heating. Roslyn feels that in a transitional period some people may be deterred by the possibility that what is now on the market will soon be obsolete. Phil replies that in 1946 right after the war automobiles were scarce. With all the wartime improvements in internal combustion engines, automobiles in 1960 were probably going to be better than the automobiles of 1946, but those advanced cars weren't going to be available for fifteen years. "People wanted and needed cars in 1946. In the same way, by 1990 improved and simpler solar systems may be available, but we need heat in the coming decade." If you can have a system keeping your house warm now, and it costs less than oil or coal, why wait?

Owning a solar home has meant a busy retirement for the Barbashes. They've visited solar and geothermal installations in Australia, New Zealand, and England. "We might have been dissatisfied with the lack of activity if we had not been participating in a venture such as this," says Phil. "We're thinking about solar energy, we're talking about it, we're fostering it." Roslyn: "We're obsessed with it."

Curves for Class

Example is the school of mankind and they will learn at no other.

Edmund Burke

This house demonstrates that a solar home need not be an ordinary-looking box with a steep roof. It can have class. Located near Aspen, Colorado, on a southern slope facing Mount Sopris and the Elk Range, it features four Trombe walls and a solar greenhouse containing a rock-filled thermal storage wall.

The original goal of builders Jed and Andra Kairath, Ron Hoffman, and Thomas Gray was to construct a prototype solar home that could compete with conventional housing in terms of cost, comfort, and style. The Kairaths were to live in the house to fine-tune it before putting it on the market. Because the crew possessed masonry skills,

The southeast side of the Kairath home. Greenhouse is in the center, with vent windows open.

architect Peter Dobrovolny came up with a brick design allowing the entire building to serve as thermal mass. Though they may be beautiful, at 60¢ apiece, plus 10¢ for transportation, the bricks were not cheap. The builders were unable to keep costs under $100,000 as they had hoped, because of the masonry and large expanses of ceramic tile.

Brick Thermal Storage Walls

The Trombe walls, brick like the rest of the house, are interspersed among large areas of glass so that warmth and light are transmitted to each room. Because the builders were tired of black, the walls were painted with a latex called "Oregon Grape." "We weren't sure," Jed said, "that we were going to like those big splotches of blue, but they turned out so well that we repeated the blue in the interior tile work." Before the glass was put in, the walls were tested to make sure the blue color would absorb enough heat. Polyethylene was stretched across them and a thermometer inserted...to discover that the air was at 140° F., just what you'd expect from a Trombe wall.

The Trombe walls radiate heat directly to the rooms. In addition, a convective loop is created as cool air from the house flows through a trap into the space between the Trombe walls and the glass, is warmed, rises, and reenters the living spaces. Says Jed: "You can hold a candle down at the bottom of the wall and watch the smoke move down and under the Trombe wall—or hold it at the top and see it move into the house. You can't really feel the air move, but you can see it."

Self-inflating Drapes

At night or on cloudy days, automated Mylar drapes prevent heat loss through the glass, thus

Front entrance on the southwest side with log steps. Stoneware planter by Jed Kairath.

TECHNICAL DATA

Owner-builders: Jed and Andra Kairath, Thomas Gray, Ron Hoffman, Colorado
Designer: Peter Dobrovolny
Solar Consultant: Ron Shore

General Features

Latitude: 38° N
Degree-days: 8,910
Insolation: 200
Heated area: 2,000 ft^2
Year of completion: 1978
Insulation: Walls: 3" foil-backed urethane; 3" fiberglass where buried
        Roof: 6" fiberglass and 6" sod
        Floors: 2" urethane
Orientation: 3° E of S
Solar system: Hybrid—passive with fan (Trombe walls)

Collection System

Passive collector/storage: 324 ft^2 (4 Trombe walls)
  Angle: 90°
  Cover: Double glazing of plate glass
  Absorber: 8" brick walls painted dark blue
  Insulating curtain: Self-inflating Mylar drapes by Thermal Technology Corp., (THERMATEC), El Jebel, CO

Active collector: 150 ft^2 greenhouse
  Angle: 90°
  Cover: Single layer of glass
  Absorber/storage: 15'×8'×20" brick wall filled with rocks
  Fan: 1/12-hp

Distribution System

Air circulates by natural convection off Trombe walls through house. Fan pulls air off top of greenhouse through rock storage and back through vent at right bottom corner of the wall. When air in greenhouse is below 85° F., storage is sealed and heat radiates to living areas.

Auxiliary System

Backup: Rumford fireplace and electric baseboard system

Domestic Hot Water

Collector: 32 ft^2 drain-down
Cover: Single layer of glass
Absorber: Copper pipes on aluminum plate
Manufacturer: THERMATEC
Backup: 30-gal. electric water heater

Costs

House: $115,000
Solar: $4,000 (materials, excluding brick)

eliminating the possibility of reverse flow. Ron Shore of Thermal Technology Corp. (THERMA-TEC) invented them in response to the need for a less expensive insulating system than the Bead-wall. Ron, who formerly worked for Zomeworks installing Beadwalls, says that Beadwalls do the job of insulating that they were intended to do, but installation is tricky: "You just can't send a Bead-wall out to somebody with instructions enclosed and expect him to put it in." So he designed the Mylar drape, half as expensive as a Beadwall when it's installed, and "idiot-proof."

In the winter when the temperature in a sensor behind the glass reaches 90° F., the curtains auto-matically ascend out of the way allowing the sun to reach the brick. Made of two layers of rip-stop nylon with a coating of aluminized Mylar, the shades, when they are down, dress the Trombe walls in gold lamé. Warm air off the walls enters through slits in the back of the material, causing the layers to separate and puff up. By means of this "inhalation system" the insulative airspace be-tween the layers can be doubled or tripled in size to attain a thermal resistance of between 8 and 12. A U-track on each side of the drapes provides an edge seal while the bottom of the shade contains a weight which rests against the floor. The seals are not very tight, however, meaning that the larger

The self-inflating Mylar curtains cut nighttime heat losses in winter and prevent solar heat gain in summer.

the curtain the higher the R factor because of a higher surface-to-edge ratio.

The necessity for providing movable insulation for glazing areas is particularly obvious in a cold climate like that of Aspen. Ron estimates that in a near-9,000 degree-day situation the drape system will have a payback period of 4.2 years (based on electricity rates of 3½¢ per kwh, and assuming a 15 percent per year increase in the cost of power). At $5 a square foot, the shades are competitive with conventional drapery, which often costs over $2.50 per square foot—more with insulating liners.

Illustrated details of curtain system (courtesy Thermal Technology Corp.).

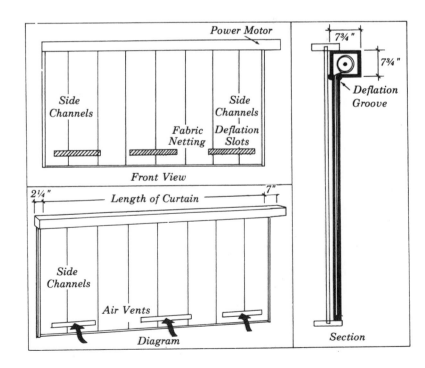

Jed reports that the curtains are easy to install and operate. When triggered by the sensor, a motor winds them up on a tube contained in a long plywood box. "You just leave a little extra room at the end for the motor and plop the box in and nail it down." Tin flashing above the Trombe walls permits access to the motor and roller for servicing. Once the motor is wired and the thermostat set, the curtains function automatically, although there is a manual override switch. From inside we found the whir of the motor audible but not objectionable.

Developing the curtains was not easy. At first there was trouble with the limiting switches that stop the curtains at the top and bottom. The original ones, crude and homemade, kept getting out of adjustment. Now Ron is using a manufactured off-the-shelf switch for reliability. The external motor is being replaced with an "in-tube" device to save space. The cost of this change resulted in the rise in price from $4 to $5 per square foot. One problem hasn't been solved: on a cloudy day the curtains go up and down frequently. The manual override permits an owner who is bothered by the

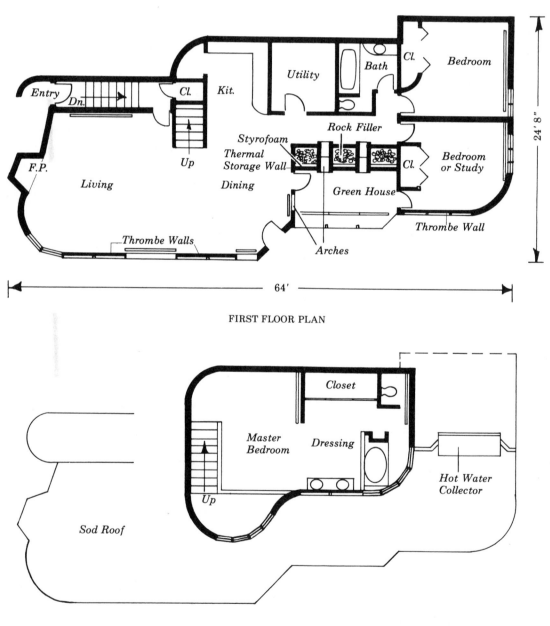

FIRST FLOOR PLAN

SECOND FLOOR PLAN

movement or concerned about wear-and-tear to fix them in any position.

Heat from Greenhouse

The greenhouse with its rock storage also supplies a major portion of the house heat load. The north (interior) wall is brick with a hollow middle filled with 8 yards of rocks. An inch of Styrofoam between the bricks and the rocks prevents heat from seeping from the storage bin into the greenhouse. Air off the top of the greenhouse circulates through the rocks and back to the greenhouse through a vent in the wall. At night a back-draft damper seals the vent, and heat gradually radiates to the hallway and the study.

During June and July the Kairaths left the Mylar curtains in the furled position to see how the structure would perform without them in hot

Tile floors are used throughout for thermal storage. All windows are double glazed except here in greenhouse, which overheats.

weather. Until the first of August, the temperature of the house stayed between 68° and 72°. As the sun moved lower in the sky, the afternoon temperature would rise to 76° or 77°. Allowing the drapes to fall reduced it to 72°. The greenhouse, despite the installation of an automatic exhaust fan, heats up to 95° F. in August. Either a larger fan or some means of shading is needed.

Backup Not Needed

After six months in the house, Jed and Andra were surprised at the efficiency of the design. Andy is a real estate agent and Jed a plumber and potter. This is the first house they have designed and built. The shell of the building was finished in March, and after the electric baseboard system brought it up to temperature—a three-day task—it stayed at 65° F. with no further help from the

Looking into the greenhouse from the dining room. Thermal storage wall at left is filled with rocks. Upstairs is the master bedroom.

auxiliary system. Occasionally during a snow-storm the crew would burn construction wastes in the fireplace, but they had to be careful to avoid overheating.

Jed laments the $2,000 tied up in the unused electric backup, which was installed only to please lending institutions. The fireplace, built in accordance with the principles of Count Rumford, is quite sufficient for auxiliary heating. Developed in the 1700s just before the advent of the Franklin stove, the Rumford design carried fireplace efficiency to new heights. Using a tapered brick-lined chimney and an external draft, the fireplace is high and shallow to throw heat out into the room. According to Andy, it warms the whole house in a half-hour, but a smoldering fire may mean some puffs of smoke in the room. Also, lacking doors, it cannot be stoked for overnight burning.

Sod Roof

Sod covering the Kairaths' roof serves an esthetic rather than a thermal pupose, since the 6-inch layer has an insulating value of something like two. "In the winter you end up with a big clump of frozen dirt," Jed says. A thicker layer would insulate better, but the roof would require more support.

Jed would not build a sod roof again on a structure this complex. Construction involved cleaning the wood decking with solvent, then putting on a membrane of butyl rubber for waterproofing. "The rubber works, but it's a mess," Jed says. "If it gets folded over, it sticks to itself and you stick to it trying to get it apart." After the roof was completed, it sprang two leaks and the sod had to be dug up. The grass also requires upkeep—watering, trimming, edging—which is difficult on a roof with so many levels and curves.

The curved outlines and the reflective curtains give the house a modernistic character while the sod roof and blue Trombe walls reintegrate it with its natural surroundings. Throughout the interior the large expanses of brick wall are interrupted with masonry arches lined with blue mosaic tiles. In the center of one Trombe wall, a surprise treat is a stained glass window revealed when the curtain rolls up. Massive beams supporting the sod roof are exposed in the living room and master bedroom. Along with handcrafted elm and oak cabinets, the kitchen has all the American conveniences from trash compactor to garbage disposal. Combining his talents, Jed took special care with the baths, using hand-thrown sinks and expanses of tile. Adjoining the master suite is a tile-lined dressing room and step-down whirlpool bath.

View of first floor's sod roof (center) and second story. Flashing protects the Trombe walls.

View from the southwest.

The features which lend the house its character also made it hard to build. It took eight months, with the four crew members working seven days a week, ten hours a day. The difficulties started with the excavation. From the surface it appeared that burying the rear wall of the house would be easy, but big rocks turned up that took an earth mover five days to remove. The curved corners were made of plywood, bent and covered with cedar battens, which went on slowly until a nail gun was rented. Inside, a circular well rising from the dining room to the master bedroom was constructed by soaking dry wall until it bent to the curve formed by the plywood plate. Laying up the brick walls and ceramic tile floors was the most time-consuming aspect of all.

Although no bank was interested in financing the house, private funding was not hard to secure. The solar era has dawned in the Roaring Fork Valley. Of twenty-five houses under construction in this particular subdivision, four are solar-heated. Ron Shore has installed Mylar curtain systems in eleven Aspen area homes, including that of singer John Denver. Jed believes that once people realize passive heating is cost competitive with conventional building techniques, solar houses will go like wildfire.

The builders are looking forward to constructing a house embodying the same heating principles as this one but on a lower budget. By making Trombe walls of concrete block with stucco or dry wall on the interior and using native lumber, they expect to be able to produce a thermally efficient house—with class—for under $40,000 in materials. Of course, a one-acre lot 40 miles from Aspen currently sells for $25,000. Truly low-cost solar housing is not going to begin here. But the kinds of innovative design being explored may provide some of the breakthroughs needed to extend solar heating to the rest of the nation.

New Life

It is myself that I remake.
W. B. Yeats

Given new life by the addition of a solar-heated wing, this typical Maine farmhouse also represents a new life for its owner—and his father. Noel Stookey (better known as Paul of Peter, Paul, and Mary) moved to Maine for the same reason that Christopher Fried (see p.63) fled to Pennsylvannia: he wanted the sanity of country life. And like Christopher, Noel bought a henhouse, but instead of raising chickens he converted the three-story barn into a recording and animation studio. A year

Southwest side of the house showing the new wing with the air collector (at right) and the original farmhouse. Noel bought the henhouse in 1973, decided then on solar heating, bought the farmhouse and built the addition in 1975. Photo by Claude Bolduc.

later he acquired a farmhouse across the road, complete with a florist business that his wife, Betty, runs, a vegetable garden, a meadow for grazing sheep, and a spectacular view of the sea. Then Noel asked his father, George Stookey, to design a solar-heating system.

Papa Stookey, as he introduced himself, has the right background. Although not a graduate engineer, he's been "at" the engineering field since 1928. Working as a heating consultant for several years, he participated in designing the first automatic coal stoker. But in 1973 when Noel approached him about solar heat, he told his son it would be completely uneconomical. "All right—I want to do it anyway" was the reply. Noel thought they could try to **make** it economical, although what mainly kept him in it was "a thirst for self-sufficiency, which was part and parcel of living in the country."

The elder Stookey is above all a practical man. What counts is how much you save on your oil bill. So he experimented on collectors for four years, trying everything for the absorber from cut-up beer cans to steel wool. He eventually designed a beautifully simple and lightweight collector that looks quite promising.

George Stookey has a new life too. "I have to credit Noel," he says, "with nudging me into something besides a retirement of entertaining and annoying myself—to something that I could get my teeth into." You may be wondering whether he would still caution Noel against solar heating. No way, not with oil prices going up.

"Fairwinds," as the Maine ocean-side house is called, was enlarged to accommodate the Stookeys and their three daughters. In the new section a formal but livable main room overlooks the bay. Walking through the original living room, which is dominated by an old upright piano, to the new wing you're hardly aware of the junction between the old and the new. A TV room, kitchen, and mudroom complete the first floor. Upstairs, the bedrooms have the casual and comfortable feel of old farmhouses.

Roof-Integrated Collector

The collector, which replaces an ordinary roof, reflects a philosophy of keeping things as simple as possible in an active system. The absorber is backed by an inch of fiberglass ductboard, which combines structural support and insulation into one material. Styrofoam was tried but it melted and popped like popcorn from the intense heat. Seventeen of these absorber-insulation units are set into the spaces between rafters.

Aluminum the weight of TV dinner trays was etched and sprayed with flat black paint, then Papa Stookey formed it into a step-like configuration using a machine of his invention. Each riser is an inch wide and set at a 30° angle. Each tread is perforated at intervals of about an inch. At the bottom the steps are three inches from the glass, then angle so as to close the whole opening at the top. Air flows in from below, entering in front of

---

TECHNICAL DATA

Owners: Noel and Betty Stookey, Maine
Solar consultant: George Stookey

General Features
Latitude: 44° 30' N
Degree-days: 7,900
Insolation: 140
Heated area: 3,800 ft^2 (2,000 ft^2 new addition)
Year of completion: 1975 (solar addition)
Insulation: Walls: 6" fiberglass and 1" Styrofoam
　　　　　　 Roof: 6" fiberglass
Orientation: 15° W of S
Solar system: Active air

Collection System
Collector: 680 ft^2 gross, 480 ft^2 net
　 Angle: 50°
　 Cover: Outer layer ¼" plate glass; inner layer ⅛" glass
　 Absorber: .005" aluminum, nonselective coating, 1" rigid ductboard

Storage System
Container: 4'×8'×21' poured concrete bin (550 ft^3)

Material: 30 tons of rocks 1"-2" diameter or eutectic salts
Location: Crawl space beneath new addition
Insulation: 1½"-2" urethane on inside

Distribution System
1/3-horsepower blower circulates 160° F. air through collector at approximately 1,500-1,600 cfm. Furnace blower takes hot air from storage through furnace where it will be further heated when necessary, and then through ducts to house.

Auxiliary System
Backup: 138,000-Btu oil furnace

Domestic Hot Water
Winter mode: preheated in copper coils laid on top of rock bin.
Summer mode: water supply piped up to heat exchanger in plenum at top of collector and heated before hot air is exhausted to outdoors.
Boosted by electric water heater.

Costs
Solar: $5,000-$7,000, plus cost of eutectic salts.

the black surface and filtering through the perforations to pass behind the aluminum and flow out into a plenum at the top. Angling funnels the air away from the glass at the point of greatest heat loss, at the top where the air is at its hottest.

In creating turbulence within an air collector, the trick is to avoid putting in so much resistance that blower horsepower must be increased. Turbulence is often obtained by adding pieces of metal

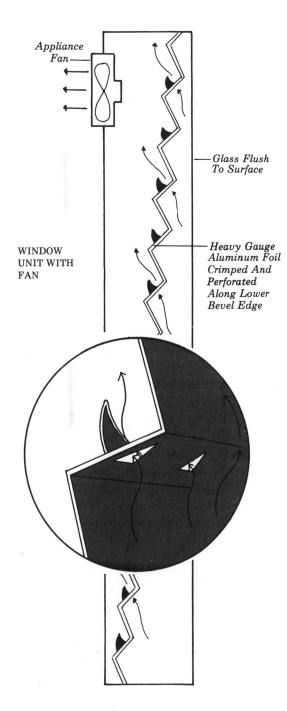

George Stookey and the back of the air collector (at right).

(fins) for baffles, but the Stookey collector has punched holes instead. Passage through the perforations created the turbulence. Simplicity: glass in place of shingles, ductboard instead of plywood plus insulation, and a one-piece absorber.

Flexible System

The collector has additional advantages. It permits a great deal of flexibility in that the tilt of the risers can be varied for different latitudes, rather than changing the whole roof pitch. At some latitudes it could even be mounted on vertical walls. For those who prefer to avoid building a steep roof, and for those who want to retrofit a house that has a less-than-ideal pitch, a design like this could solve their problem better than ordinary flat-plate collectors.

The weight of the blackened aluminum, only a pound per 12 square feet, is another plus—especially for retrofits. Moreover, since the absorber is fabricated in one piece, mass production could be simplified. Machinery might even be designed to adjust for custom-producing any given tilt.

The inventor of this collector believes it's a misconception to think of solar heat as a total heating system: it should be thought of as an aid. Disagreeing with the Saunders outlook (see p.56), Papa Stookey believes that trying for enough storage to get all your heat needs from the sun becomes un-

106

economical. "You put about two-thirds more into it to get that last 50 percent."

Instead, people should think of solar heating in stages: first add a collector to heat the house during the daylight hours, then later add storage.

Generally speaking, a house is kept hotter during the day and the thermostat is turned down at bedtime. Thus a system that gives daytime heat is saving just as much as one that has storage for nights, at maybe half the cost. At any rate it's important, George urges, to get started on whatever you can afford. To do otherwise is like spending your life waiting to find out who you are before you start doing anything. If you delay until you can afford a complete system, you'll miss out on savings that you should have now.

Stressing that the collector was custom-made for Noel's home, George points out that it is not commercially available. But if his present ideas pan out, a spin-off is possible in the form of a lightweight, inexpensive, portable panel that could be installed in windows like a shutter, so that people could start taking advantage of the sun in a small way. He describes it as a peanut concept: "We're talking about peanuts in the way of heat, and I don't know how practical it will be." A small appliance blower mounted in the back of the heat-absorbing shutter would circulate hot air around

Noel "Paul" Stookey was on a tight schedule because Peter, Paul, and Mary were rehearsing for a reunion concert tour and albums.

New living room and dining area looking toward old living room and kitchen. Photo by Claude Bolduc

107

the house or apartment. With lightweight materials—the Stookey absorber, plastic for a cover, and insulation for the structural part—the unit could be hinged like a casement or slid on a track. It could even be put up and taken down like a storm window. Heat loss reduction would go hand-in-hand with heat collection.

Two Closed Systems

Consistent with the philosophy of simplicity, once Noel's collector is finished, heat collection and distribution will operate as two closed systems. When solar heat is available at the collector, a blower control will trigger a fan to send air through the collection system. A backdraft damper in the feed duct and one in the return will prevent cooling of the storage through reverse circulation at night. When the house thermostat calls for heat, it will switch on the furnace blower just as it does now. But ducts will route the return air through rock storage, then the heated air will flow through the furnace and back into the rooms. If the solar-heated air is not warm enough, a control in the furnace will start the oil burner. There will be no separate mode for bypassing storage to heat the house on sunny days. When heat is needed, the house will get it from storage whether or not the collector is running.

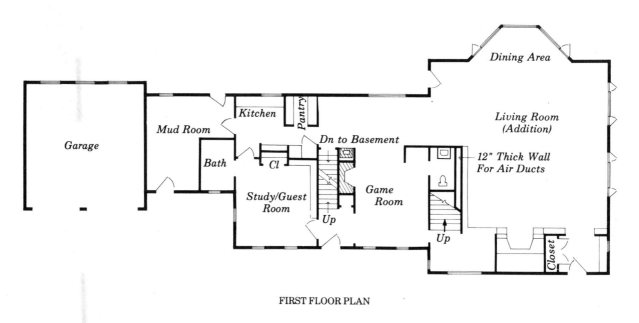

FIRST FLOOR PLAN

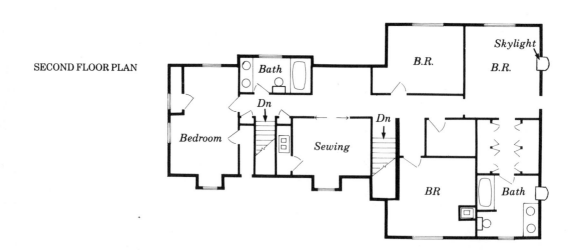

SECOND FLOOR PLAN

Southeast side with the florist business in the background. Photo by Claude Bolduc.

For domestic hot water a tiny circulating pump will be needed for summer operation when the water to be heated will be piped up to a heat exchanger in a plenum at the top of the collector. No pump will be needed in winter because the water will be heated in copper coils laid on top of the rock bin. The storage tank will be above the bin to take advantage of the fact that water rises when heated. The same tank will be used for the summer mode, so the changeover will involve simply closing one valve and opening another twice a year.

Costs

Costs are relatively low for an active system. The integral construction of the collector together with the simplicity of its design are the main reasons, but it helped, too, to find materials that could be recycled, like a used blower. So did pouring a storage bin at the same time as the foundation for the new wing. Most of the ductwork is not included in Noel's cost estimate because ducts for the new wing had to be installed anyway.

Glass was the most expensive item. The outer layer was put up in 4-foot × 10-foot sections. For such large spans, plate glass was required since snow loads can be heavy in Maine. Angle iron, seated on a rubber cushion and silicon strip to allow for expansion, was screwed down to put pressure on the glass to hold it in place. The glass company failed to anticipate the temperatures involved. Their mastic sealing material melted and allowed some glass to slide down a half-inch. Stops were installed to prevent further sagging, but the gap can let in rain.

The gross area of collection surface was reduced by three view windows inset into the collector. Air is blown across these narrow skylights just as it is with the rest of the collector, the only difference being that there is no black absorber. In addition to loss from the windows, the net area was less than expected because of faulty carpentry work that meant rafters had to be shimmed, and because of the addition of extra rafters that the architect felt were needed. Since the roof angle, designed to harmonize with the original house, is low enough for the June sun to overheat the collector, an exhaust fan was installed and an expansion-type control may be added to open vents in case of a power failure.

Noel and his family have lived in the house since the new wing was added, even though the solar heating was incomplete. Recently the ductwork was finished so that next winter the sun should be making a contribution. At present a blower in the storage bin is pulling the solar-heated air into the unfilled bin. The bin, not yet closed off, heats the basement to 75° F. George predicts that once the rock bin is filled, the system will provide between 20 and 40 percent of the heat load.

For now he's holding off on rocks and seriously considering using salt hydrates to increase storage capacity. He recently visited the University of Delaware and learned that eutectic salts could provide twelve to seventeen times more heat storage than a comparable volume of rocks. A hundred cubic feet would store one million Btu. He is checking into the possibility of building a walk-in storage with the salts stored in containers. He believes a seven-day storage is feasible. In the meantime he's already looking ahead to attaching vertical collectors to the recording studio.

Would Change System

Noel describes the system he would use if he were to build again: "Design-wise the roof is the wrong place to put your panels." To preserve normal roof lines and window areas, he would attach the collector to a daylight basement. With the storage nearby, the heated air could be returned by natural convection. Noel might even go further and eliminate air collectors in favor of covering the basement wall with glass to make a Trombe wall, with a concrete slab for additional storage. Vents to the upstairs would complete a totally passive system.

Neither Noel nor his father sees the price of commercially produced active systems coming down significantly. Materials are likely to go up as long as the price of energy to produce them goes up. Even if mass production were to lower costs, the savings would not be passed on to consumers so long as manufacturers charge what the market will bear. As the price of petroleum increases, people will need and want collectors more, and be willing to pay more, so the companies will lower prices only enough to encourage more people to buy.

Noel is pessimistic about the role of business in solar heating: "I think this is going to be one of those few enterprises that will have a very successful underground life. It's created, promulgated, and passed on by word of mouth, and people are going to be inspired by each other. This is a grass-roots thing."

The Bugs Are Mainly Microscopic

Please don't take my sunshine away.

Song

Although officially retired from his position as professor of microbiology at the University of Rhode Island, Dr. Philip Carpenter is making a unique contribution to his field by monitoring the bacterial count in his solar heating system. Incubating cultures in a laboratory in his basement, he's found that the count in his storage tank tends to fluctuate but is not high, varying from 18,000 bacteria per milliliter of water (a teaspoonful) to as low as 5,000.

In summer, the tank temperature gets up to 130° F., which is hot enough to kill most microorganisms. The remainder are ordinary water bacteria, nothing to pose a health hazard. Phil did

---

The southwest view. A solenoid valve drains this collector system when the temperature of the panels drops below 50° F. If the electricity should fail, the valve opens automatically.

isolate a curious population of thermophilic, or heat-loving, bacteria that has never before been described. "There's probably not an awful lot for them to feed on in there," he says, "which limits the population. They grow and die and the next generation lives on the decomposition products of the first, keeping a steady state population."

No Clogging

When he initiated these studies, he was mainly concerned about the possibility of slime formation, since algae in quantity could clog pipes. He has seen no evidence of this, however, and believes that the counts are low enough that it will never become a problem. Only in a trickle collector where the water is exposed to sunlight, favoring the growth of one-celled organisms, could algae conceivably increase to the point of blocking the feeder pipe.

Acorn, the company that manufactures precut solar homes, provides a 6-ounce bottle of algaecide at the time of installation to kill any accumulated growth already in the tank. (See p. 50.) Although the Acorn collectors have copper tubing so that water is not directly exposed to the sun, Acorn's maintenance schedule includes examining for algae twice a year at the same time that the water level is checked. The rest of their maintenance schedule is the same as Phil performs on his system: once a month he cleans the air filter on the blower and once a year he adds two or three drops of oil to it.

Collector Shaded

The home of Phil and Helen Carpenter, a Rhode Island shorefront cottage, stands on the crest of a hill where a scattering of large oaks only slightly deflects the ocean gales. Phil, using an instrument that measures solar radiation, estimates that in winter they lose 10 to 15 percent from shading by the oaks. Unfortunately some are on their neighbor's property, the only case we found of an incipient problem with "sun rights." Helen mentions a nearby subdivision in which the houses, designed for adding solar panels at a later date, were deliberately situated to avoid interfering with each other's sunlight.

Siting was particularly important with the Carpenter home. The main floor is even with the top of the hill while the basement is dug into the side of the slope facing the sea, effectively earth-berming three walls of the lower level. Containing the lab, a family room, and workshop, the basement is unheated except for spillover from the heat pump and solar storage. Even so, the temperature has never fallen below 62° F., partly because the basement is largely underground and also due to the floor construction: 5½ inches of concrete poured over gravel, then 2 inches of Styrofoam and 4½ inches of additional concrete. With all this plus carpeting

---

TECHNICAL DATA

Owners: Helen and Philip Carpenter, Rhode Island
Designer: Spencer Dickinson

General Features
Latitude: 41° 30' N
Degree-days: 5,800
Insolation: 140
Heated area; 2,400 ft^2
Year of completion: 1976
Insulation: Walls: 6" fiberglass and polyethylene vapor
                barrier
        Roof: 12" fiberglass
        Floor/foundation: 4"-6" fiberglass/2" Styrofoam
        Shutters: Insulated drapes on all windows
Orientation: 15° E of S
Solar system: Active liquid drain-down

Collection System
Collector: 360 ft^2 (fifteen 6'×4' panels)
    Manufacturer: Solar Homes, Inc., Jamestown, RI
    Angle: 57°
    Cover: Double layer 0.040" Kalwall Sun-Lite®
    Absorber: 16 oz. copper sheet with ½" diameter copper
        tubes soldered 6" apart on centers, nonselective coating,
        6" fiberglass
    Pump: ⅛ hp

Storage System
Container: 8'×22'×5' poured concrete tank with 5 coats
    epoxy
Material: 4,500 gal. water
Location: NE corner of basement
Insulation: 4" Styrofoam (bottom), 6" fiberglass (top and 2
    sides), 1" Styrofoam (other sides)

Distribution System
Thermostat triggers ½-hp pump to send water from storage to a 20-gal. electric water heater (for keeping input to heat pump above 45° F.), then through 44,000-Btu heat pump. 1/3-hp blower in heat pump sends 120° F. air through ducts to rooms.

Auxiliary System
Backup: Heat pump and 20-gal. water heater and fireplace
Fuel consumed: 2512 kwh at 4 cents/kwh average and 2
    cords at $25/cord

Domestic Hot Water
Preheated in coil immersed in storage tank, boosted by heat pump, then by 80-gal. electric water heater.

Costs
House: $65,000
Solar: $6,700

Southeast view of Carpenter home looking up from boat dock.

with a rubber liner, trips to the basement in stocking feet are comfortable even in the dead of winter. Six inches of fiberglass insulate the basement ceiling from the main floor, although Phil now believes this was unnecessary. Having seen how well the lower level performs, he regards underground housing as quite promising.

The Carpenters' collector was manufactured by Solar Homes, Inc., a construction company owned by Spencer Dickinson. Spencer designed Phil and Helen's home from their sketches, built the collectors, and supervised construction. He also services the system, making him one of the pioneers in getting all the elements together. If anything goes wrong, Helen calls him and he's there in twenty minutes. Having received this kind of individual attention, she and Phil quite naturally advise prospective solar homeowners to find a reliable local solar builder and then trust in him. The Carpenters found Spencer through a solar home show.

One-stop service helped keep solar costs down although the overall price tag on the house went considerably over the original estimate. The Carpenters attribute this to galloping material costs at the time they built, plus their tendency to pick quality materials when it came time to choose items like tile, tubs, and windows. As the bills

mounted, Spencer felt so bad that he decided to pass up his profit margin, while Phil helped by doing some of the finish work.

The solar heating system cost $3,000 for the panels, $1,000 for the tank, and $2,700 for the heat pump package. Ductwork added another $1,300— but ducts are needed in any heating system. The heat pump package, which included both the back-up and domestic hot water heaters, today costs around $4,500.

System Monitored

Equipment on loan for a year from the University of Rhode Island measures input and output temperatures from the solar collector. During the first winter the tank temperature stayed above 53° F. and during the second year above 67° F. The back-up heater has never gone on, since its only function is to maintain a 45° input to the heat pump. The Carpenter home has operated exclusively on the solar-assisted heat pump plus a living room fireplace.

Heat from the fireplace is fed into storage by means of a system of Phil's design. A pair of pipes lead from the tank to a coil welded behind a steel

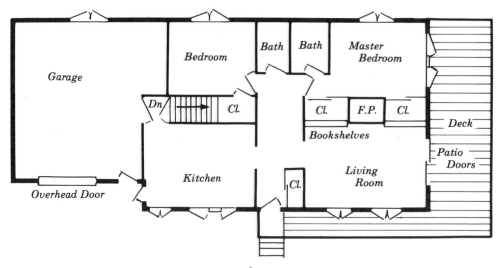

FIRST FLOOR PLAN

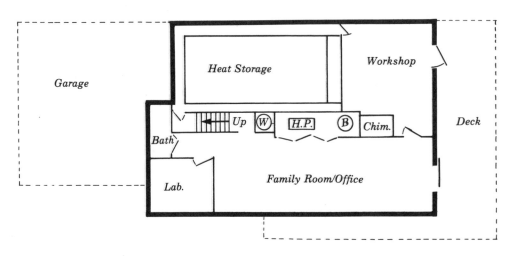

BASEMENT PLAN

plate at the rear of the firebox. When a manually operated switch is thrown, a 1/12-horsepower pump takes water from the tank at a depth of 30 inches, circulates it through the coil, and returns it at a depth of 21 inches. In February of 1978 the fireplace contributed one and one-half million Btu; 3 million were contributed by the solar panels. In January the situation was reversed since it was a cloudy month. The Carpenters used the fireplace nearly every day in December, January, and February, but by mid-March were almost finished with it. Phil is so pleased that he wishes he had increased the number of pipes by running a coil up the chimney flue.

The pumps, water heaters, heat pump, controls, and monitoring equipment are all compactly contained in a closet. Data carefully logged in notebooks indicate that the Carpenters used $140 worth of electricity plus 2 cords of wood in their first winter for space heat and domestic hot water. This was reduced in the following year to $112, because of milder weather and improvements such as an insulated steel door for the kitchen entrance. With $50 for wood in 4-foot lengths, which Phil saws and splits, the second year's fuel bill totaled $162. This included the blower and all other operating costs. A couple of years ago the local electric company borrowed the Carpenters' house plans and did a heat loss study. The result was an estimate of $385 per season for resistance heating. Even using that figure, which sounds rather low to us, the Carpenters save $223 a year at the electric rates of two years ago, plus whatever they save on hot water. They plan to live long enough to see their investment amortized.

Not having a dishwasher, which requires hotter water, the Carpenters set their water heater at 120° F. When we were there in mid-March, the solar storage was 90° F., necessitating a 30° boost by the heat pump and water heater—in June the heat pump can do it alone. Summer bills for water heating run $8 a month, which Phil calculates is a savings of 71 percent.

Air Conditioning

During hot weather the Carpenters can set the thermostat to reverse the heat pump to provide air conditioning. Excess heat is taken from the house and dumped into the solar storage. The first summer Phil ran the collector to see how hot the water would get. When air conditioning was needed, the tank was too warm to accept heat from the house. The heat pump switched off automatically to avoid pressure buildup. This summer the Carpenters may turn the collector off so that the air conditioning can perform properly. But at the same time this would eliminate solar water heating. Alternatively, when the season for air conditioning approaches, the Carpenters could run the solar panel at night and dissipate some of the heat through the roof, but this again would suspend domestic hot water heating. Another possibility, which would call for additional expenditures, would be to install an outdoor tank and run a pipe to duct the heat out. Phil is not overly worried about finding a solution, because air conditioning is only needed two or three days each summer.

Other problems encountered have been minor. Extra coats of waterproofing had to be added to the concrete storage tank when leaks developed. Insulation on the tank had to be increased because the room containing the tank was heating up to 95°. Afterwards it dropped 15°, while the tank temperature ran consistently higher. The only other bug that needed correcting was a faulty compressor on the heat pump that gave out within a year but was replaced under warranty.

At night and on cloudy days the Carpenters faithfully close the foam-backed curtains that Helen sewed, except for the living room drapes

Kitchen. The house receives a fair amount of direct-gain solar heat.

facing their boat dock. They leave those open while they eat supper watching the reflections of sunset on the bay. Their thermostat is set at 68° F. during the day although passively gained heat sometimes warms the rooms to 74°. Helen likes to think of the house as "scooping up the sun." At night they set the thermostat back to 62° F. and close off the master bedroom, which drops as low as 58°. The guest bedroom is kept closed most of the time.

"It's the most comfortable house I have ever lived in," Helen says. No drafts and a pleasant humidity level are two reasons for her enthusiasm. Her plants bloom year-round, thriving on pure air and sunshine, and the house seems to stay cleaner. Even her silver needs less polishing.

Both Carpenters are sold on solar heating. They are disturbed that their local electric company, after installing several solar hot water systems, put out negative reports. Apparently some of the systems were unsuccessful, but the Carpenters find this difficult to understand—given that theirs performs well. One week after a representative from the utility gave a talk to the local Rotary Club, Phil presented a slide show on his house. The two talks created such opposite impressions that the Rotarians no doubt wondered whom to believe.

The Carpenter house, with its cedar shingling, blends with its neighbors so well that houseguests and delivery men often fail to realize it's a solar home, even though the collector faces the street. Phil reports that people drive in to make deliveries, return to their truck, look back at the house and say, "Hey, is this a solar home?" A neighbor child dropped by after school one day and said he'd just made a discovery in science class: "You people probably don't know it, but there isn't another house like this on our street."

Owning a solar home has meant an active retirement for Phil, what with bacterial investigations, slide shows, and meticulous record-keeping. In this he's contributing grassroots research on solar-heating systems and much-needed information on how solar homes work out in actual practice.

Dwellings in the Sun

Here, in thirty-two pages of color photographs, are the answers to many questions about solar heating. Do solar homes "look different?" They can, but they need not. Are they expensive? Some are; many cost little more than comparable non-solar homes. What kind of people live in them? All kinds. Old and young. Rich and poor. Most of them are concerned about energy waste, try to cut unneces

Wolcott-Lindsey House

Nine inches of clinkers insulate roof from heat of summer, cold of winter. At right, reflector doors increase solar gain by at least one-third. Below, Kiva measures corn in greenhouse.

Davis House Collector and double-glazed windows soak in Maine sunshine.

Ramage House Robert Ramage uses small tractor to spread stones once used to store heat.

Van Winkle House   In summer vegetables grow both inside and outside this greenhouse, a part of the Van Winkle house. At right, Rip addresses visitors, using public address system.

lcomb House Two-story expanse of glass
closes balcony plus area for plants. Below, adobe
eplace in living room is part of backup system.

Sanford House   An active collector, double-glazed windows and Sarah Sanford, the owner, vie for sunshine on a wintery day. Below, right, wood paneling adds warmth to bath on north side of house. Center photo courtesy Claude Bolduc.

Ritter House A solar home model that's on the market. Windows tie dining room to outdoors.

Saunders House   In contrast to this, the north side, the south side is a two-story bank of windows, with the bottom level set at an 80° angle.

Fried House This retrofit greenhouse provides heat plus year-round vegetables.

Zaugg Houses Two solar methods are compared in these homes. Ranch house at left has water system, while A-frame has air collector. Photos below show north side and living room, kitchen of A-frame.

Nystrom House   Sandstone cliff dwarfs this three-story structure. Inside, Southwest Indian motif is carried out in bedroom and other rooms.

Barbash House   The aptly named Sunship, its evacuated tube collector and its salon.

Kairath House Four Trombe walls and greenhouse are the solar system for this house. Interior view shows door from dining room to greenhouse.

Stookey House   A New England farmhouse is comfortable with its solar system retrofitted. Wing on left side of house was added. Right, the living room in the new wing.

Carpenter House A hilltop site, a view of the lake, and heat from a system that combines solar and a heat pump.

Pendell House
Double glass windows and doors offer sunny living for a retired couple.

Gilbert House   Tracking collectors eye the
moving sun to heat this house, with its view of the
Raggeds and the Uncompahgre Mountains.

Caivano House   Roc, Katie and Helen Caivano frolic in the sunlight. Dining area, too, is bathed in sunshine.

Bunn House
A bank of collectors on the
south side; windows and
balconies are seen from
the northeast.

Johnson House A salt-box with collector.

Maxwell House   An arc of twelve collector panels.

Curtis House

Ingredients: aluminum cans
and a greenhouse.

Cohen House   Redwood adds beauty to this home, both inside and out.

143

Sigstedt House   The solar trailer, with
its additions, includes (top photo) the
diorama window of the greenhouse; (right)
the greenhouse, and (below) the kitchen
in the trailer.

Rusher House   Big windows, a comfortable interior—and a new use for a garage.

Jantzen House Circular windows resemble eyes in this house. Views are of south side, at top, and southwest corner. Below, what it's like to peer into one of the eyes, into studio inside.

Jantzen House   In top photo, light pours into
bedroom from two sides and top. Below, the northwest
corner of the house.

A Simple Solar Retirement Home

Do not build your house upon sand. Built it upon rock. Lots of little rocks.

Unknown

When Claude and Vivian Pendell retired to their cattle ranch in sunny eastern Washington state, they wanted to build a simple house requiring little maintenance. They had just returned from Indonesia where Claude had been an advisor on sanitizing slaughterhouses. They tried living in an apartment in Spokane but hated it. They had seen their friends who bought retirement homes and then were saddled with big monthly payments and high maintenance costs that made a trial of years that should be comfortable. So they decided to design and build their own solar-heated home.

They built it for $18 a square foot, including furnishings, well, and septic system, and in the process they came up with some innovative solutions to typical owner-builder problems.

Driving out from Medical Lake after a snowfall to visit the Pendells' underground structure, we

found ourselves going around in circles, looking in vain for the house. A pickup truck surrounded by dogs rounded the top of the hill, and we hailed the driver to ask for directions. It turned out to be the Pendells' daughter, Patrice, and her friend Robert Johnson, out running their sled dogs in preparation for a competition. We followed the sleds to the house, which we had overlooked because it is so unobtrusive.

Built into the side of a south-facing mound overlooking two small lakes, the house collects solar heat passively through two sliding glass doors and double-glazed windows in the living room. Heat is stored in a gravel bed underneath the floors. The wood floors stop two feet short of the south wall of glass so as to expose a narrow strip of gravel that serves as a "heat register" as well as a drain for Vivian's plant collection. A fan pulls air through the gravel to the floors in the north rooms, thus spreading the solar heat through the house. The Pendells vent the air out the back of the house, but by relatively simple alterations to the system it could be recirculated into the rear rooms if a higher degree of efficiency were desired. In summer the gravel floors help cool the house.

Troughs Not Needed

Troughs in the gravel facilitate the flow of air, but Claude now feels they are unnecessary and may even reduce the storage of heat. The plywood flooring is supported by 1×4's and 1×8's laid flat on top of the gravel. These "joists" were arranged like baffles to insure that warm air reaches every part of the floor. Claude took the idea of building on gravel from the railroads: if trains can run on ties sitting on gravel, why not build a house that way? But, showing us a spot where the floor is springy, he emphasizes the importance of tamping the rocks thoroughly.

An advantage of this method is that it's cheaper than wood framing although it might not be appropriate in all locations. "Using gravel was possible here," says Patrice, "because the eastern Washington soil has natural gravel deposits from receding glaciers." Drainage is excellent. One problem

is that this type of flooring does not conform to insurance regulations, so the Pendells have been unable to get household coverage.

Temperature Swings

Using their living room rather like a greenhouse full of exotic plants, the Pendells are not overly concerned about heating efficiency and feel no need for covering their glass at night with drapes or shutters. "Why, Claude even opens the bedroom window at night, no matter how cold it gets," Vivian exclaims. "He says it's better for your health to sleep in a cold room." When they wake in the morning, they reheat the house if necessary with one or more of three electric heaters located strategically in the bedroom, living room, and bath. Including the cost of their all-electric kitchen, washer and dryer, and small kiln, their electricity bills average $25 a month during most of the year, rising to as much as $50 in January and February. In case electric rates go up too much they've provided a stovepipe opening for installing a wood stove in the living room.

We found that, without drapes, the living room does become quite warm during the day, especially near the windows where most of the plants are located. The Pendells, as they grow older, appreci-

---

TECHNICAL DATA

Owner-designer-builders: Claude and Vivian Pendell, Washington

General Features
Latitude: 47° 30' N
Degree-days: 8,000
Insolation: 110
Heated area: 1,000 ft^2
Year of completion: 1976
Insulation: Walls: House is earth-integrated; 8" concrete walls with 1½" airspace
Roof: Wood, 10" earth
Floor: Gravel
Orientation: S
Solar system: Hybrid—passive with fan

Collection System
Collector: 100 ft^2 of double glass windows and doors
Angle: 90°

Storage System
Material: 8" gravel
Location: Subfloor

Distribution System
Fan pulls warm air through exposed gravel bed at foot of glass doors. Troughs in gravel facilitate flow of air to rear of house, where it is vented outside.

Auxiliary System
Backup: 3 electric heaters
Fuel consumed: $25 in January

Costs
House: $18,000 including furnishings
Solar: Negligible

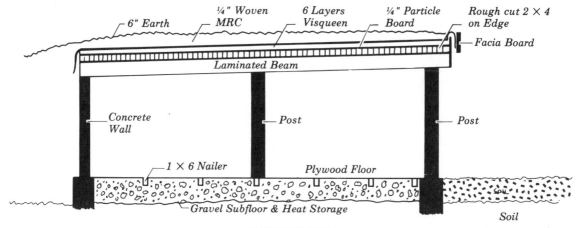

6" Earth · ¼" Woven MRC · 6 Layers Visqueen · ¼" Particle Board · Rough cut 2 × 4 on Edge · Facia Board · Laminated Beam · Concrete Wall · Post · Post · 1 × 6 Nailer · Plywood Floor · Gravel Subfloor & Heat Storage · Soil · Soil

EAST ELEVATION

ate the warmth more than do their visiting children. Furniture, which might be faded by the sun, is arranged against the interior wall facing the view. If the room becomes uncomfortably hot, one of the sliding doors is opened a crack. The sliders are used as a front entrance, but the laundry with its side door serves as a winter air lock entrance and receptacle for boots and coats. The gravel "register," another barrier against mud on the carpet, can simply be hosed down.

Claude and Vivian did most of the construction work themselves with the aide of Patrice and Robert, a few carpenters, a plumber, and an electrician. As ranchers, they had built many outbuildings as well as a passively solar-heated house

in the 50's. The retirement house went up fast; after the concrete walls were poured in March, 1976, it took only ten actual workdays to complete the shell. There were delays in finishing the structure—weather problems, recurring "Indonesian intestines," but by July the Pendells were established in their new home.

Earthen Roof

Constructing an earthen roof was one of the trickiest problems they encountered. Unsure as to how thick to make the roof supports to hold the weight of the sod, they sought advice from a local factory

Schematic of heating system.

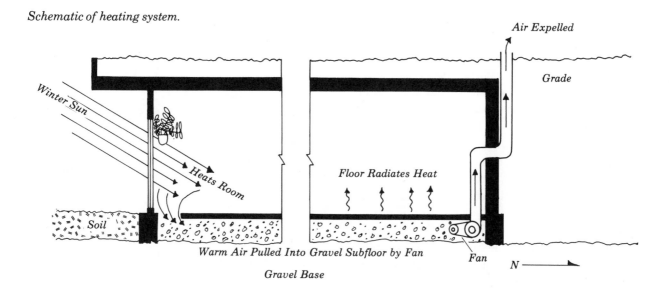

Winter Sun · Heats Room · Soil · Warm Air Pulled Into Gravel Subfloor by Fan · Gravel Base · Floor Radiates Heat · Air Expelled · Grade · Fan · N ⟶

Barnwood was used for exterior siding and interior paneling.

that makes laminated beams and were told to use 4×12's on 6-inch centers. Although supposedly no central support was required, Claude added two posts anyway as a precaution.

The ceiling is composed of a solid layer of rough-cut 2×4's placed on edge across the top of the beams. Besides offering strength, this approach created an attractive ceiling, which Vivian's stained-glass hanging lamps highlight. The 2×4's are roofed with particle board to prevent splinters from puncturing the six layers of 6-mil plastic laid on top to prevent leaks. After a 4-inch cushion of earth, metal mesh ("hardware cloth") was laid next as protection against burrowing rodents. The Pendells' suspicions about the need for a barrier were confirmed when they were visited last summer by a mole who emerged from the stovepipe opening, having dug in from the side of the roof, where there's no mesh.

Finally, 6 more inches of earth were spread over the roof. After becoming saturated during the winter, the dirt settled a little, requiring that more be added before seeding with grass and poppies. The strength of the roof was tested accidentally by one of the workhorses that Claude raises. A mare wandered up on top of the house while the earth was still wet and panicked because she thought she was trapped in the mud. She stomped around a lot before escaping, but did no damage. Claude still

feels that he should have put up more roof supports, so that he wouldn't have to worry about such things.

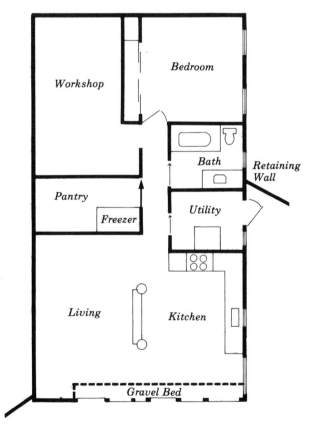

PENDELL FLOOR PLAN

Vivian acted as contractor during the building process, organizing the purchasing end of the operation. She was also the interior designer. Rather than buy Sheetrock, the Pendells tore down an old barn on the ranch and used the wood to cover the concrete walls. Sealing the concrete would only cause moisture to condense on the inside, Claude thought, and therefore the walls were left to "breathe" naturally. The concrete was painted black so that cracks between the barn boards would be invisible. The boards were nailed over strips of furring, creating an airspace for insulation.

Although friends advised them not to use dark barn wood throughout the house, they did so anyway and are happy with the rustic effect. Windows just below the roof line in the bedroom, studio, and bath help to lighten the northernmost rooms. Because the kitchen is in the southeast corner, it catches early morning light. One of Vivian's ideas was to brighten the kitchen by repainting her refrigerator. She took it and the dishwasher to an automobile paint shop and had them sprayed red-orange, thus saving money and adding color.

Comfortable House

The Pendells find this the most comfortable house they have lived in. It fulfills their hope for easy maintenance: no painting of the barn wood, interior or exterior, is required (but there are some big windows to wash). Vivian wishes the house were a bit larger. Although she has a studio for her stained-glass and painting projects, she has no place to lay out her sewing. Also, they may put a

Vivian Pendell did the interior decorating and stained glass work.

window in the only spot in the south wall that isn't already glass, to capture even more sun. Originally Vivian thought she might like to have one corner for privacy in the living room but has come to see no need for it.

The house has some good features for a retirement home that the Pendells hadn't consciously considered: all the rooms are on one floor and the sliding doors provide large entranceways for walkers or wheelchairs. Best of all, it is paid for out of savings and the owners can look forward to using their retirement income for other purposes. The house could be adapted for a less sunny or more severe climate through the addition of wall insulation, drapes or shutters, and a heat recovery system.

Concentrating on Collection

Modern definition of a sundial: the thermostat on a solar-heated house.

B. C. Cartoon

Because they wanted baseboard hot water heating, Harry and Peggy Gilbert set out to find the most efficient system for gathering the sun's rays. They liked the hot water system in their old farmhouse and saw no reason to change when they began designing their new home a few hundred yards down the road. As owners and operators of a Colorado plumbing and heating business, they had access to research materials on solar heating, but it was not easy to find a collector that would turn out the 140° F. water required in a baseboard system. Finally, after two years of reading and attending seminars, Harry ran across the Northrup concentrating collector in the office of a plumb-

ing supplier in Denver and was convinced he should try it.

In the meantime the house was constructed and outfitted in preparation for installing a solar heating system. Built of heat-absorbing slump block from New Mexico, it was well insulated and equipped with a storage tank in the basement. Since Harry was unsure what system he would use, he started out big: he had three compartments, each holding 1,200 gallons, cast in concrete in the mechanical room. The baseboard system was installed in three zones regulated by separate thermostats.

For the first year the Gilberts heated with their

154

auxiliary electric boiler and read the electric meter daily, testing heat load calculations. The baseboard radiation system had been sized 30 percent larger than usual because of the relatively low solar water temperatures. As it turned out, the house did better that first year than was predicted, largely because Harry had neglected to consider the factor of passive gain through the front windows. "On sunny days the house temperature will get up to 80° with no help whatsoever, even though it's only 20° outside," Harry says. "The natural passive heat was a wonderful surprise." The Gilberts had thought that much heat would be lost through the windows, but because of the view of the Raggeds and the Uncompahgre Mountains, they decided to make the sacrifice. Actually studies have shown that properly oriented double-glazed windows gain more heat during the day than they lose at night, even in severe climates. The net gain is small, however, unless shutters are used.

After four months the passive system began to reveal drawbacks: the carpet faded and as the low winter sun poured in, morning glare became so intense that Peggy had to discontinue holding committee meetings at her house. Her mother

Harry, a plumber, assembled the collectors and built the hot water baseboard heating system. Here he checks the electric wiring.

made drapes for the dining room windows but they shut out too much light. At last the addition of a reflective film to the south-facing glass stopped the fabric deterioration and reduced the glare. The film, with a blue tint, reflects or absorbs 80 percent of the ultraviolet radiation while cutting radiant heat losses, at a cost of about a dollar a square foot.

TECHNICAL DATA

Owner-designers: Harry and Peggy Gilbert, Colorado

General Features

Latitude: 39° N
Degree-days: 6,000
Insolation: 210
Heated area: 3,940 ft^2
Year of completion: 1977
Insulation: Walls and roof: 6" fiberglass
　　　　　　Foundation: 2" Styrofoam
Orientation: S
Solar system: Hybrid—active liquid with antifreeze and
　　passive direct-gain

Collection System

Passive collector: 143 ft^2 (double-glazed windows)
Active collector: 280 ft^2 (28 concentrating tracking collectors)
　Manufacturer: Northrup, Inc., Hutchins, TX
　Angle: 45°
　Cover: Acrylic fresnel lenses
　Absorber: Slightly flattened 1" copper pipe, selective coating, 3½" fiberglass
　Pump: ½ hp

Storage System

Container: Concrete tank
Material: 1,200 gal. water
Location: NW corner of lower level
Insulation: 2" Styrofoam (walls and top)

Distribution System

Water from collector warms storage water through heat exchanger. A pump triggered by thermostat circulates water through baseboard hydronic system.

Auxiliary System

Backup: Electric boiler and 2 fireplaces
Fuel consumed: $199 at 3½ cents/kwh average

Domestic Hot Water

Copper coil in storage preheats water for 50-gal. electric heater

Costs

House: $125,000
Solar: $12,000 (materials, wholesale)

But since the film also cuts heat gain, it has marginal applicability in solar installations.

The passive heat helps the Gilberts get by with twenty-eight concentrating collectors instead of the thirty-eight they would need if they were trying to get all their heat from the active system. Buying another 100 square feet of collectors would have added $3,000 to the cost of the system, making it noncompetitive with factory-made flat-plate panels. Although concentrating collectors are more expensive per square foot than most ordinary collectors, Harry is happy with his choice: "It's been fun and more interesting than something plain. But I've had to do quite a bit of adjusting and playing with the system."

The Northrup collectors were surprisingly easy to assemble. Harry thought he would need help putting the kit together but managed to do it by himself in the evenings after work. Of course, his BS degree in mechanical engineering helped. But anyone who could install a baseboard hot water system could also assemble the solar components, although a system this complex would not ordinarily be considered a do-it-yourself project.

Follow the Sun

The Northrup collectors follow the sun across the sky from sunrise to sunset, providing 30 percent more efficient conversion of solar radiation than flat plates, the manufacturer claims. Sun-tracking is accomplished by attaching the collectors to a cable and pulley drive mechanism. By turning a worm screw, a small high-torque electric motor can drive up to twenty collectors attached to the cable. The motor is controlled by two photovoltaic cells mounted on the pulley shaft. These silicon cells, situated at a 60° angle to one another, must be placed in full sunlight. When one cell receives more light than the other, the difference in current generated triggers the motor, which turns the col-

Acryllic Fresnel-type lenses cover each collector panel. Below is the Gunnison River, and in the distance, the Raggeds.

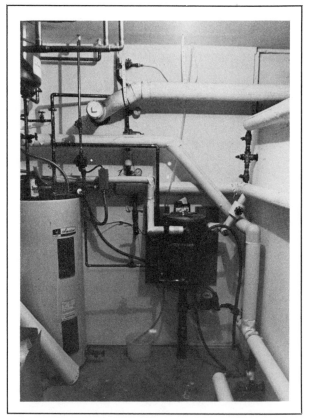

Storage tank on right, 42-gallon electric water heater on left. Above it is 25-kw electric boiler that boosts solar system. Black tank at center is heat exchanger (collector liquid to water).

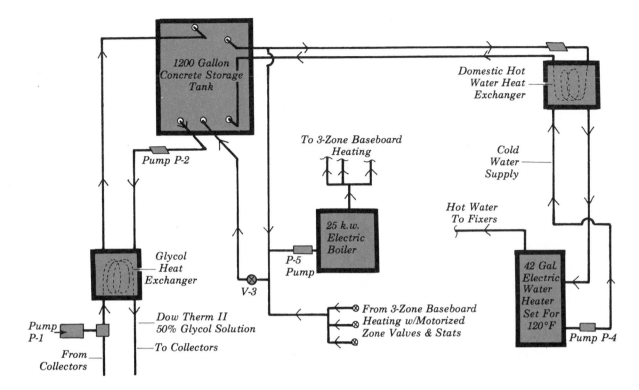

Schematic of Gilbert solar system.

lectors in the appropriate direction until equilibrium is restored.

High performance is achieved through the use of curved fresnel-type lenses one foot wide and ten feet long covering each collector. Made of extruded ultraviolet-resistant acrylic, the smooth outer surface of the lens transmits radiation with a minimum of reflective losses. As the sun's rays penetrate the lens, they are refracted by a computer-designed array of prisms that concentrate them on a copper pipe containing water.

Problems

The collectors may be set at varying tilts from the horizontal, but the manufacturer recommends the angle of latitude. Harry instead installed his at 45°, not too far from his latitude of 39°, but easier to handle. He aligned the collectors in two banks of fourteen each, with two drive motors. "They say you can get by with twenty on one motor," he says, "but I've had trouble with the motors burning out within six months as it is." His supplier, far away in Denver, has afforded little help in debugging the system, and Northrup, Inc. has been under-

going an identity crisis for the past year, as Atlantic-Richfield was attempting to buy the company. Harry wrote to Northrup about the motor problem but after six months has still received no reply. "The little fellas only cost $15, but I don't want the system stopping twice every year," he says. Presumably once the corporate struggle is over the company will be more responsive.

The perplexing question is how to get the collectors to track concurrently. High winds tend to dislocate individual collector panels, which prevents them from operating efficiently. If a lens is not aimed accurately, the focal point misses the pipe and all the effort is in vain. At present, Harry checks the array frequently and puts renegade collectors back on course by hand. Additional screws on the cable may help.

Other Hassles Minor

Other hassles have been minor. Last summer some of the swivel fittings on the manifold pipes began to seep a little. Wiggling them around was sufficient to stop the drip. "They don't leak much," Harry points out, "but when you pay $5 a gallon for antifreeze, you can't afford to let too much of it fall on the ground."

Another way to lose antifreeze is by boiling it away. In summer the temperature of water in the collectors is kept below the boiling point by a mechanism that stops the tracking motors when the water temperature reaches 150° F. This is one advantage tracking collectors have over flat plates, which gather heat constantly unless covered. During the winter the pump forces water through the collectors fast enough to prevent it from exceeding 180° F. A battery-started gasoline generator keeps the pumps running in the event of a power failure. One day, however, Harry had been working on the system and forgot to turn the pump back on. When he came home from work the next day, he found that the liquid had flashed to steam and he had lost 10 gallons of solution (5 gallons of antifreeze) and melted one sensor.

The antifreeze is an ethylene glycol compound made for solar collectors. It is toxic but Harry has designed the storage system so that there would have to be a sequential breakdown in two systems in order for the domestic water to be contaminated. Although the chemical has additives which prevent it from deteriorating under high temperatures, the additives must be replenished every few years if the entire system is not to be drained and refilled. After a year, it's time for Harry to send a sample to a lab to see how his solution is doing. He also intends to check the water in the baseboard system for algae and corrosion, the latter because he has never analyzed the chemical composition of the water, especially its pH.

Storage Too Big

"I discovered the expensive way how much storage I needed," Harry admits. His 3,600-gallon tank might have worked if it had been well enough insulated. Styrofoam was glued to the walls but not the floor, because Harry incorrectly assumed

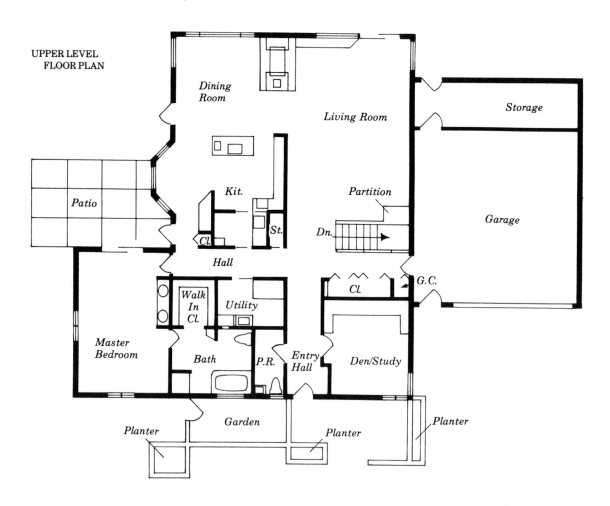

UPPER LEVEL FLOOR PLAN

Direct gain provides much of heat, although it is reduced by a reflective film added to cut glare. The fieldstone double fireplace is used mainly for atmosphere even though it has a heat return and outside air intake.

Gilbert bath with blue enamel tub. Fence outside provides privacy.

he would lose little heat out the bottom. Because it proved impossible to maintain 135° in all three sections of the tank, only the first section is now used. It was insulated with 6 inches of fiberglass on the partition wall and an inch of high-density Styrofoam plus 6 inches of fiberglass on top. This tank consistently maintains temperatures above 140° F. Harry has since encountered a rule of thumb for water storage: 3 gallons of storage for each square foot of collector. Although he no longer uses four times what is needed, his tank is still a bit large by this measure.

The first winter the system was tested turned out to be unusually cloudy. During January and February only 40 percent of the days were sunny, rather than the normal 60 percent for western Colorado. Consequently the Gilberts paid $199 for supplemental heating for the winter season. But that looks good when compared to the previous year's $600—especially for a spacious 4,000 square-foot home. Even during that cloudy winter, the Gilberts met 66 percent of their space heating needs by collecting sunbeams.

The biggest changes in their life-style brought by solar energy, Peggy says, are that Harry de-

votes much of his free time working on the system, and their social life has picked up. But Harry is quick to emphasize that they don't drop everything to give house tours, what with the plumbing business to run, farming 50 acres, and raising a few horses.

Realizing that his system is too expensive and too complicated to be applicable in most residential situations, he looks forward to the day when solar heating—especially passive—will be incorporated into the design of all public buildings: "If there's one thing I've learned from this experiment, it's the importance of passive solar heating."

During our visit we discovered something about concentrating collectors that even the Gilberts hadn't realized. It occurred to us that a photograph taken from inside the array might be interesting, so we inserted head and camera between the fifth and sixth collectors. Our shadow fell on the sensor and the motor started the apparatus tracking. We retreated faster than a scared turtle backing into its shell. There was really no danger of getting caught, but it was a disconcerting experience, although one which only photographers need worry about.

A Hobbit Lair

A preference for round windows, and even round doors, was the chief remaining peculiarity of hobbit-architecture.

J. R. R. Tolkien

Although neither its doors nor its windows are round, the Caivano home feels like a hobbit dwelling. Perhaps it's the arched roof or the mushroom-like overhang of the second floor. Or perhaps it's the front door, which is 8 inches lower than standard doors, evoking a sense of entering a special shelter. Probably most of all it's the owners themselves, Roc and Helen Caivano, friendly, hospitable, and somehow whimsical like a couple of hobbits.

For fun, Roc used to buy oversized mayonnaise

Caivano home southwest corner. The west wall is wood frame with cedar siding on both levels. The south wall is a two-story Trombe wall enclosed in a Plexiglas greenhouse. The north wall is concrete block on the lower level, the shingle roof on the second floor. Snow bounces sunlight onto the Trombe wall, increasing its heat collection.

jars and soup cans to put on glass shelves in a farmhouse where he and Helen once lived. "You'd walk into the kitchen and feel like a mouse because the big cans were out of scale. It felt like Alice in Wonderland." Roc, an architect who has designed and built twenty-five homes, says that his house would never do in New York. But a hobbit house fits the northern Maine locale where the Caivanos live.

Psychologically speaking, the curved roof makes it easier to stay in a small space and its structural strength means you can get away with less material. Although a sphere provides maximum volume for surface area, domes felt claustrophobic to Roc, so he made the dome concept linear by turning it into a vaulted roof. The result is a home that nestles into the land, appearing smaller and lower than it really is, with a naturalness of contour that blends well with the surrounding meadow and nearby hills. The house has received such favorable national attention that Roc was one of nineteen designers invited by the American Institute of Architects Research Corporation to submit plans for a passive solar home.

When the Caivanos were first considering solar heating, Helen felt she couldn't handle the idea of an elaborate mechanical system. She and Roc, who prefers hand tools to power machinery, do not consider themselves mechanically oriented people. After all, Tolkien tells us that hobbits, though skilled with tools, neither understand nor like machines more complicated than a hand loom. A passive collector's freedom from maintenance appealed to the Caivanos. They decided on a south-facing greenhouse enclosing a sliding glass door and a Trombe wall with a fan, a hybrid system.

Fan on Trombe Wall

In conventional passive versions of a Trombe collector, the air heated on the surface rises by natural convection and enters the room through a vent.

In the Caivano Trombe a thermostat kicks on a fan when the wall reaches a high enough temperature. The fan forces the heated air down the wall and through ducts embedded in the floor, thus moving it down low where it's really needed. Roc calls it a "hot air radiant heat system," similar to the heat distribution systems used by the ancient Romans and rural Koreans. Since the blower and Plexiglas cover were installed not long ago, that part of the system is pretty much untested. Until recently solar heat was collected by opening the sliding-glass door to the greenhouse that encloses the Trombe wall. Now that the fan is in, the Caivanos report that the floor is comfortable to walk on in stocking feet.

Sinking ducts in concrete can be a tricky business. Luckily, not long before the slab was to be poured, a plumber happened to mention that ducts must be tied down to keep them from floating like inner tubes. So Roc wired them every five feet, but even so when the cement truck started pouring, one began to surface. Roc ran to hold it down, standing there with his fingers in the concrete until the duct stayed in place. (He got his hand out in time.)

TECHNICAL DATA

Owner-designers: Roc and Helen Caivano, Maine

General Features

Latitude: 44° 30' N
Degree-days: 7,900
Insolation: 140
Heated area: 1,300 ft^2
Year of completion: 1976
Insulation: Walls: 3" sprayed polyurethane foam
          Roof: 9½" fiberglass
          Foundation: 3" urethane and 2" Styrofoam (both
                  extending out from perimeter)
          Shutters: 1" fiberfill comforters on S, W, and E
                  sliding glass doors
Orientation: S
Solar system: Hybrid—passive with fan (Trombe wall glazed
    with Plexiglas and then enclosed in Plexiglas greenhouse)

Collection System

Collector: 168 ft^2 (Trombe wall); 384 ft^2 (glazing on
   greenhouse)
  Angle: 90°

Cover: Single layer Plexiglas spaced 4" from Trombe wall
Absorber: 14' × 12' × 8" Trombe wall (concrete block filled
    with gravel, painted dark brown)

Storage System

Material: Concrete block Trombe wall, 6" concrete slab on
   top of 2' gravel bed, and concrete block walls on N and E

Distribution System

Trombe wall charged during day radiates heat to interior at
night; triggered by a differential thermostat, a 1/3-horse-
power blower draws air down surface of the Trombe through
6" ducts embedded half in the gravel bed and half in the slab,
and then to registers in the living areas.

Auxiliary System

Backup: 20,000-Btu wood stove (chimney within Trombe wall)
   and one electric wall heater
Fuel consumed: 3½ cords at $40/cord

Costs

House: $30,000
Solar: $887

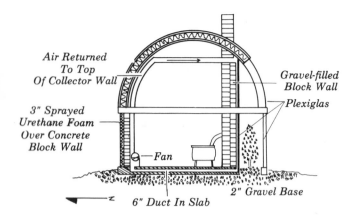

Air Returned
To Top
Of Collector Wall

3" Sprayed
Urethane Foam
Over Concrete
Block Wall

Fan

Gravel-filled
Block Wall

Plexiglas

2" Gravel Base

6" Duct In Slab

Section through Caivano house, showing air system.

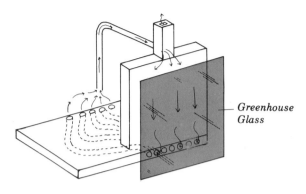

Greenhouse
Glass

Detail of Trombe wall and floor ducts.

With a two-story thermal storage wall, one-story block walls, and a concrete slab, the house has a great deal of mass. The walls were coated with Block Bond™, a glue-like substance composed of cement impregnated with glass fiber. Colored either gray or white, it resembles stucco. On interior walls it passes for old-fashioned horsehair plaster.

Polyurethane foam was sprayed on the exterior of the block, allowing the concrete to be accessible

Helen, Roc, and Katie Caivano in front of the greenhouse.

for storing heat. The insulation was then coated with another layer of Block Bond to protect it from weather. The second layer also serves a safety function since this type of foam gives off cyanide gas when it burns. It was safe to use on the Caivano house only because it's sandwiched between two fireproof layers: concrete block walls and a quarter-inch layer of bond. For esthetic reasons Roc is gradually facing the exterior with hand-cut fieldstone.

Block bonding is ideal for self-help building: Helen and Roc did the outermost coat with the help of his parents. Helen assured us that it's as easy as icing a cake, but Roc suggests starting in a non-critical area to get the hang of it.

Since Block Bond is impervious to vapor, weather surfaces need not be ventilated, which means a reduction in heat loss. Ordinarily there would also be less wind infiltration with this type of construction, but the Caivano home is buffeted by winds so fierce that a 50-pound stone used as a spark arrestor was torn off the chimney. It landed in the back-yard.

Heat-Saving Features

Heat-saving features in the Caivano house include the heavy insulation, double glazing, closets on the north wall, insulated drapes fabricated from fiberfill comforters for the sliding glass doors, and a custom-designed exterior door containing an inch of urethane insulation. The chimney, besides being centrally located, is enclosed within the Trombe wall so that heat from the wood stove is stored in the masonry. Air channels within the wall direct heat to the bedrooms. Since the blower for the thermal storage wall is operated only while the sun is shining, the use of peak hour electricity is minimized. In addition, the clothes drier is vented inside to help heat the rooms, and Helen often postpones running it until the heat is needed. For the Caivanos the changeover to solar heating involves more than a technological innovation.

Helen and Roc find such changes a delight. "It's never a burden to open the fiberfill quilts," says Roc. He likens living in a solar home to sailing a boat. The insulated quilts are opened or closed a little at a time depending on the sun and wind, just like sails being trimmed. Helen appreciates the feeling of greater control over their energy supplies: "It's like the difference between going into the mountains on a tour with a guide where everything's done for you, versus doing it yourself. We're doing it for ourselves."

The Caivanos find they can get through a twenty-four-hour period when it's sunny without using any backup; at night if the outside temperature dips below zero, the house will be down to the mid-50s by morning. During the day they aim for a temperature of 68° F., but this is hard to control. Although the Caivanos find handling 3½ cords of wood a winter no inconvenience, using wood meant finding someone to come in to stoke the wood stove when they went away at Christmas for three weeks. An auxiliary electric heater was in-

Roc in the greenhouse with the Trombe wall to the left behind another layer of Plexiglas spaced 4 inches out from the wall.

stalled in the bath for preventing frozen pipes, but a housesitter was still deemed necessary—mostly to keep the house warm for the Caivanos' cat. The electric heater is used so rarely that they see no noticeable increase in their electric bill.

Possible Changes

Helen believes that the house works so well that few changes could enhance it. An air lock for the front door and a larger greenhouse that could be used as a sitting room are improvements she suggests. Roc wishes he had designed the house to catch more early morning light and had insulated under the concrete floor. The question of insulating slabs involves a trade-off between heat lost by conduction—meaning an uncomfortably chilly floor—and thermal mass gained by the ground being close to the heat source—meaning protec-

Wood stove in living room in front of Trombe wall.

tion from freezing for the house. Roc also advises anyone trying this system to use a thicker Trombe wall, 12 to 16 inches instead of 8. In the late afternoon of a sunny day the wall will be "very warm," by midnight "a lot cooler," and by morning "very cool." Roc believes the radiant heat would last much longer if the wall were thicker.

Recent research on conventional Trombe walls indicates that their functioning as thermal storage walls may be more important than their collecting function. That is, the heat radiated through the wall may be more significant than that convected through slots. Other researchers disagree. Like many solar issues, this one is still unresolved.

Helen's only serious complaint concerns the performance of the house in springtime when Maine weather is often wet but not cold enough to justify

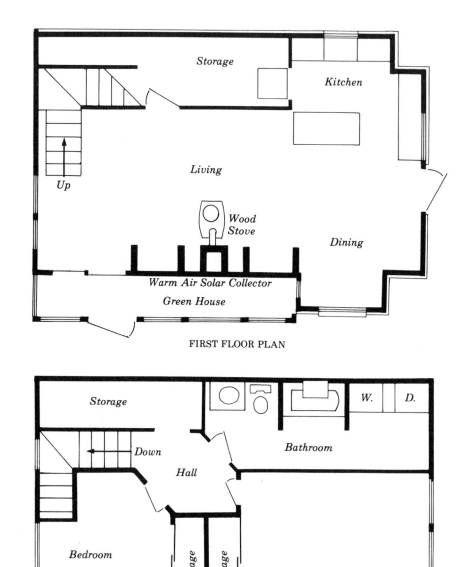

FIRST FLOOR PLAN

SECOND FLOOR PLAN

Kitchen of the Caivano home.

a fire. Even when the weather is clear, the spring sun is too high for adequate solar gain with a vertical collector. The house is somewhat dark and dank for about a month, but Helen notes that her dissatisfaction may be just a matter of spring restlessness: it's too late for winter sports and too soon for gardening. "Maybe it's just my bad time of year, so the house doesn't work for me then."

When doing the initial design, Roc found it difficult to locate reliable technical information: "I couldn't find an engineer in the state of Maine who could give me a straight answer." Even now, he says, in 1979 and probably into the early 1980's, people have to realize that they're pretty much on their own when they're designing a solar house, "and not rely on some unproven theory from a half-educated professional." Like Roc, you have to go on your own enthusiasm, which only comes from trying something you believe in. During con-

struction he was told that figures on the heat-charging rates for Trombe walls indicated that his wouldn't work. Since he was three-fourths of the way through the house, there was nothing he could do except say, "I didn't hear that." Later, after living in the house awhile, he learned that the figures had been changed.

A teacher of environmental design at College of the Atlantic, Roc is convinced of the importance of educating people about solar energy, especially the next generation. Even our preschoolers should be taught about the importance of energy conservation and solar heating. When the Caivanos' daughter, Katie Starlight, was a year and a half, she held her hand where the sun was hitting the floor and said, "See—hot," and then put it in the shadow, "See—cold." When Katie grows up, no doubt she'll build her own solar-heated hobbit home.

Payback: Plus or Minus 20 to 70 Years

You pays your money and you takes your choice.
Punch

Anne and George Bunn of Madison, Wisconsin, paid their money and chose what someone once called the Cadillac of the solar heating business: a Solaron collector. They paid $11,400 for the panels and controls, $2,000 for solar ductwork, $500 for stones, and a little something for a crane to hoist the collectors to the roof. Round it off for a total of $14,000.

During the winter of 1976–77 the Bunns averaged $44 a month for gas in the four coldest months. The fuel bill for the season totaled $245, just over half the cost of heating a well-built Madison house of the same size. Thus George and Anne save roughly $200 a year on space heating plus a few dollars for hot water. With a savings of $200,

or perhaps $300 including hot water, their unmortgaged investment will take forty-six to seventy years to amortize. Fuel escalation might bring that down to twenty years before long.

As George remembers it, when they planned the house he figured on a thirty-year amortization and counted on fuel cost escalation to reduce it to twenty-five. He was surprised when we got our calculator and came up with seventy years. Either way, it's a long time.

George Bunn, a professor of law at the University of Wisconsin, believes that solar heating will never take off as long as it's financed with conventional methods. Most people will balk if solar collectors add $10,000 to $14,000 to the house cost.

George would like to see utilities own the equipment and rent it, the government provide some form of subsidy, or the banks reduce lending rates for solar systems. One alternative he doesn't mention is the people choosing less expensive systems.

More Than Saving

At any rate he wasn't motivated by the idea of savings: "I could have invested the money and gotten more out of it. We were building a house anyway. It just sounded like a crazy and interesting thing to do." Anne, a registered nurse whose first husband was an architect, has built quite a few homes but says the solar heating was George's kick. "He's enthusiastic about new things. I think he did it to blaze the way—he's that type." Anne, an active backpacker, skier, and birdwatcher, considers herself a conservationist and that entered into her decision to go ahead despite the expense.

Because Farrington Daniels, one of the grandfathers of solar energy, taught at the University of Wisconsin, the school has had a solar research lab for many years. In 1975 a friend at the lab tried to talk George out of building a solar home, because he felt that it wasn't ready for residential use, with the panels not yet off-the-shelf items as they are today. At an engineering exposition, however, George saw a domestic hot water collector producing scalding water on a cold day. He was hooked. The friend capitulated and recommended Solaron as the only company in those days that had been building solar panels long enough to give any assurance that the equipment would work as advertised.

Although George is certain he would select solar heating again if he had the choice, he's not sure whether he would recommend it to others. He admits encouraging his brother a little. Anne is much more positive: "You shouldn't postpone things in life."

TECHNICAL DATA

Owners: Anne and George Bunn, Wisconsin
Designer: Charles O. Matcham, Jr.

General Features
Latitude: 43° N
Degree-days: 7,800
Insolation: 115
Heated area: 1,900 ft^2
Year of completion: 1976
Insulation: Walls: 3½" fiberglass and plastic vapor barrier
　　　　　Roof: 10"-20" fiberglass
　　　　　Foundation: 2" Styrofoam
Orientation: S
Solar system: Active air

Collection System
Collector: 565 ft^2 (29 panels, each 6½'×3')
　Manufacturer: Solaron Corporation, Denver, CO
　Angle: 58°
　Cover: Double layer ⅛" tempered glass hermetically sealed
　Absorber: 20-oz. steel sheet with integral fins and ⅝" channels on back, nonselective coating, 3½" fiberglass

Storage System
Container: 22'×5'×5' poured concrete bin
Material: 22 tons 1½"-3" diameter rocks
Location: Subbasement
Insulation: 2" Styrofoam on inside

Distribution System
Forced-air in five modes: 1. A ¾-hp blower circulates air upward in absorber channels at 1,130 cfm., down through ducts to rock bin. 2. Another ¾-hp fan (in furnace) circulates air from rocks to ducts to house. 3. Gas furnace comes on when room temperature drops 2 degrees below rock storage temperature. 4. If the thermostat calls for heat on sunny days, dampers send hot air directly from collector through furnace fan to rooms and back to collector. 5. In summer, exterior air enters through a vent, cools rocks, and exits. During the day, house operates under mode 2 for cooling room air.

Auxiliary System
Backup: 110,000-Btu gas furnace and two wood stoves
Fuel consumed: $185 at 26½¢ per 100 ft^3 and 1 cord at $60/cord

Domestic Hot Water
Small pump circulates water through heat exchanger to 80-gal. tank. In winter boosted by 60-gal. gas water heater. In summer, gas heater bypassed (collector air just blown across heat exchanger, not into rocks).

Costs
House: $120,000
Solar: $14,000
Maintenance costs: $100 for a damper motor, $40 for a fuse

Nestled amid conifers and deciduous trees on the shores of Lake Mendota, the Bunn home is narrow like a row house. Because of the shape of the lot, the Bunns had to build up rather than out. The whole house is only 25 feet wide but has six levels. Basically it's three stories with half-stories between the main levels, which jut prow-like toward the lake. Two stories are below grade to keep the house from appearing ungainly.

Starting our tour at the bottom, George showed us the rock storage and mechanical room, which share a subbasement level. The basement, up a few steps, has a billiards room, bath, laundry, and George's study. Up several more steps the main entrance hall is on the third level along with a one-car garage. The stairwell continues with a bright Navaho rug overhead and large south-facing windows on the right. A built-in window garden filled with tomatoes, parsley, lettuce, chives, and houseplants welcomes the visitor, and some passive solar heating filters through the vegetation.

The kitchen, which George thinks of as the bridge of the house, is on the fourth level. Next to the kitchen, the living room–dining area contains

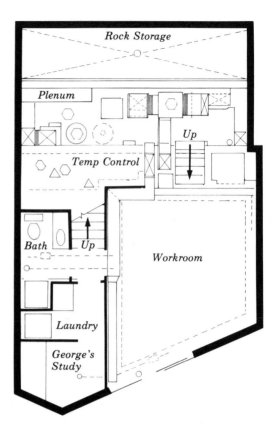

BASEMENT FLOOR PLAN

an airtight wood stove for reducing use of the gas furnace. The fifth level, with two bedrooms and a bath, is the kids' space. Matthew, the Bunns' teenage son, enjoys a room with a cathedral ceiling and sleeping loft. Daughter Sara's room, used for guests now that she's out on her own, has an attic overhead for storage. On the sixth level the master bedroom suite consists of a bedroom, Anne's study, dressing room, and bath with fixtures scaled for George's height. Skylights on the north roof of the suite are necessary for natural light since the south face of the house is almost completely taken up with solar collectors.

Nightly Heat Loss

Although George, like Anne, is an outdoors person —with him it's whitewater canoeing—he complains that Anne is a fresh-air fiend. She likes to sleep with the sliding glass door in their bedroom wide open even when it's minus 28° outside. By morning the room is below freezing. It's not sleeping in a refrigerator that bothers George as much as the fuel waste. Another wood stove heats the

The north side.

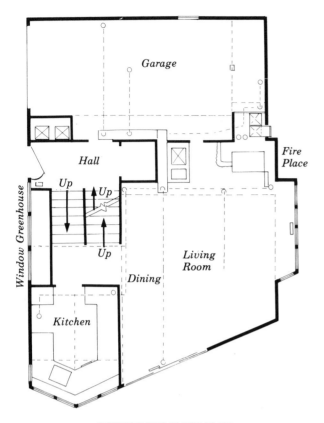

LIVING LEVEL FLOOR PLAN

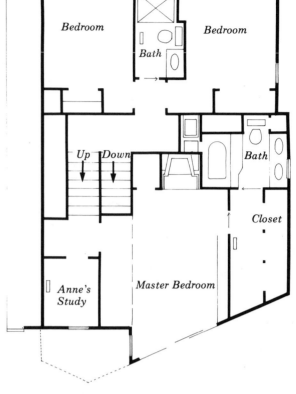

BEDROOM LEVEL FLOOR PLAN

room whenever they want to read in bed or watch Masterpiece Theater. Meanwhile in the rest of the house the thermostat at first was kept at 68° F., but solar heating eventually inspired the Bunns to set it back to 65° F.

Esthetically the height of the house is no problem since horizontal redwood siding keeps the building visually tied to the ground. But the height does mean more exposure to winter storms that blow in from the northeast off the lake. A bay window on the north side of the living room and east-facing sliders on two levels are no help. Windows are minimized on the south because it is the street side and has the collectors. The west side, facing neighbors, is mostly blank wall for privacy, but the east has windows to capture the view and the north has quite a few for light. All but the sliders are triple-glazed: nevertheless the house seems to have too much glass in the wrong places.

Improvements

Other improvements: Anne suggests that their house should have a vestibule—she finds the word

"air lock" too ominous—for the front entrance. They found it necessary to add insulation over the rock storage area to control heat loss into the garage above it. George regrets not building with 2×12 studs on the north and east walls so as to fit in a foot of insulation. But in the Bunns' case, shutters on those north windows would do more good than 12 inches of fiberglass in the walls. George thinks "something should be done about electricians" since the first thing they did was cut through the plastic vapor seal for the wall plugs and switches—causing infiltration.

The Bunns are quite satisfied with the performance of their panels. The manufacturer, Solaron, performed four inspections during construction, the first being to supervise selection of the rocks. Their representative rejected two or three lots before settling on river-bottom-rounded gravel. Most of the smallest pebbles were washed out, leaving fairly large stones.

Solaron sent another consultant to supervise installation of the collectors and another just before the ductwork was hooked up. A final check was made after the system was running. "We were

169

George and Anne Bunn. George became hooked on solar heating when he saw a collector producing water so hot it scalded his fingers on a 40° F. day.

their first installation in this part of the country," Anne says, "so that may be why they came, but we had the impression that they cared very much that their system worked properly."

Solaron has stood behind the equipment. They replaced without charge some aluminum bridge strips that discolored, a purely esthetic defect. The strips cover the cracks between panels to keep out rain. George and Anne have had to pay a couple of maintenance charges in their first two years of heating with the sun. After the first season George had a man out to check over the system, and he replaced a motorized damper that was still working but didn't close far enough. The Bunns were charged $100 for this. Then they blew a fuse, which George replaced three times. They finally called in a repair man to replace it, and it hasn't blown since. "That cost me $40," says George. "Apparently the fuses I bought weren't very good."

Savings on Gas

At the beginning of the second year they had a couple of valves installed to bypass the gas water heater in summer. If they have three cloudy days in a row, they turn it back on, something of a chore

and a bit more frightening to Anne than a gas oven but she's become accustomed to it. Because the gas company charges a monthly minimum of $5, it's hard for the Bunns to tell how much they're saving. With the gas heater off and laundry hung on a line, they use no gas at all yet still pay the minimum. Anne complains that the solar-heated water, although quite warm enough for bathing, is not hot enough for the dishwasher. Soap and food scraps are left. "I would rather wash dishes by

Dining area and see-through kitchen. Some of this glass is triple glazed, the rest double glazed.

hand, but I'm outvoted so we run the dishwasher—and I wash them by hand when everyone else is gone."

She also points out that they don't use the cooling system because it has never worked very well. If they had hot days and cool nights, and kept the house shut up during the day, it would function better. Since Anne prefers not to live in a sealed-up house, they use an attic fan and she and George sleep out on their deck on hot, muggy nights. That should be fun when the stars are out. Although breezes off the lake help, Sara—visiting for the day—says theirs is the only Madison house she knows of without air conditioning.

The Bunns had another problem, this one with homeowners' insurance. The company charged them ordinary rates but required a $500 deductible clause in case of glass breakage. Anne notes they've had no trouble with small boys or sonic booms. George adds that he doesn't know what hail as big as golf balls might do, but tempered glass is supposed to be strong.

As a lawyer, George maintains that the issue of sun rights which we encountered at the Carpenter house is a question of weighing values. Common law gives you some rights to the space directly over your property, but not necessarily toward the sun. It does not give you a right to cut your neighbor's trees. Value judgments are involved: are those trees more important than having the sun shine on your panels? Once society makes a decision about that, George says it will be easy for lawyers to draw up a statute. A knottier situation would be that of a neighbor who obtains zoning for a tall apartment building, thereby making his land more valuable. Or the fellow next door who wants to build an addition that would shade your collector. In many cases present residential zoning regulations would indirectly protect your access to sunlight. In new subdivisions careful planning plus restrictive covenants could provide adequate protection. The problem will be more difficult to resolve in urban areas.

Despite the initial cost the Bunns are happy with solar heating. George says that he's a sun lover anyway, but when he hears the blower go on and knows the sun is warming the panels, he likes the feeling of getting something "for free." Anne adds that living in a solar home is an active pleasure, partly the satisfaction of feeling noble and virtuous, the way you feel when you hang out laundry. She also finds it comforting to sit in her study—right up under the collector—and hear the roof begin to talk about what it's feeling. As the panels expand and contract, they go pwong, ping, pwong. "When you're alone in the house, you're not really alone."

Living room. Anne believes that the size of the ducts for the air collector cut into the closet space somewhat.

Off the (South) Wall
Solar Heating

A cat dozing in the sunny doorway of a barn knows all about it.
Why can't man learn?

E. B. White

"After calculating our heat load, the manufacturer guaranteed us a 50 percent reduction in our heating bill, and that's what we got," says Linda Kehoe, new owner of a Solar Room™. With additional storage the Kehoes hope to provide up to 80 percent of their heat through sun power. Their 39-foot-long greenhouse, which came in an easy-to-assemble kit, also adds 275 square feet of living space to their "big box," as they call their tract home on the edge of Santa Fe. They look forward to the day when they can afford a custom-built solar home, but in the meantime they have opted for this inexpensive, simple method of retrofitting.

Representing the "damn-the-engineering approach" to solar heating, the Solar Room is for those who balk at doing the complicated engineering analyses required for sizing collectors, pumps, and blowers. The Kehoes—Tom, public relations representative for a petroleum company, and Linda, manager of Bruce King's campaign for governor of New Mexico—are busy people. Although they are firm believers in energy conservation, they don't have the time to do careful research, so the casual approach is suited to their needs.

Solar Rooms have been on the market for two years. It took their creator, Steve Kenin, five years to develop the design and manufacturing process. With a background in psychology and engineering, he moved to New Mexico with his wife, Jean, to do engineering work for Steve Baer of Zomeworks. After working for a while with Skylids and Beadwalls, the Kenins found their interest lay more with the mass market than in highly individualized systems. Steve feels the challenge lies in retrofitting the 50 million single-family dwellings across the country consuming fossil fuels. Many are heated with natural gas, which is undependable. Steve asks: "When the crunch is on what are these people going to do?" The Solar Room was developed as a glazing system that could be tacked on almost any home to provide a substantial amount of heat. The fact that it also serves as a place to grow plants all winter long makes it even more useful.

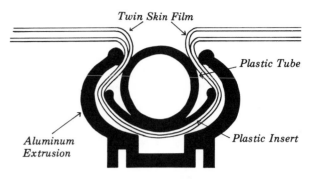

A *unique aluminum extrusion of two locking parts extends completely around the perimeter, clamping the polyethylene and sealing it for inflation.*

Inflated Skin

How does a Solar Room differ from any ordinary greenhouse? Twin-Skin™ glazing of two layers of ultraviolet-resistant polyethylene are forced apart by a static pressure blower, forming an insulating airspace that retards heat loss at night. Tension created by the pressure between the layers compresses and locks into place the arched struts gripped by an aluminum extrusion so that the structure will stand up under wet snow and high wind conditions.

Long and narrow, the greenhouse is well proportioned for solar heating in the ratio of glazing to floor space. On the other hand, the narrow width of the room limits the living space. "What happens," Steve says, "is that people buy them as greenhouses and then discover how nice it is inside and want to move in." It's possible to get custom-made versions, but adding width also reduces efficiency.

The Kehoes have already discovered the livability of their solar addition. Their original intention was to use it to start bedding plants for their garden, but the children, Brian, 2, and Larry, 10, liked to play in the greenhouse during cold weather. A sandbox was installed but there is not quite

---

TECHNICAL DATA

Owners: Linda and Larry Kehoe, New Mexico
Solar Designer: Stephen Kenin

General Features
Latitude: 36° N
Degree-days: 6,000
Insolation: 250
Heated area: 2,000 ft^2
Year of completion: 1978 (retrofit)
Insulation: Walls: 3½" fiberglass
        Roof: 6" fiberglass
Orientation: S
Solar system: Hybrid—passive with blower

Collection System
Collector: 7'×39' solar greenhouse
  Manufacturer: Solar Room Co., Taos, NM
  Angle: Variable curve
  Cover: Twin Skin™ of ultraviolet-resistant polyethylene
    inflated by blower
  Blower: 16-35 watt

Storage System
Containers: 9 55-gal. black steel drums
Material: Water
Location: Against S wall of house

Distribution System
Air flows by natural convention through doors from the greenhouse to the living room and through a window to the bedroom.

Auxiliary System
Backup: Natural gas forced-air furnace
Fuel consumed: $20 in February, 1978

Costs
House: $40,000
Solar: $1,300

enough room to fit in a breakfast table and chairs, as Linda would like, and still allow space for a walkway.

The Solar Room, according to Steve, is easy enough for "anybody's grandmother" to put up. This may be a bit of an exaggeration, but while it can take two weeks to build a greenhouse from scratch, the Solar Room can be erected in two hours to two days by two people with a screwdriver and a drill. We helped set up a demonstration model for a Sun Day exhibition in less than three hours. Thus the system is oriented more toward the urban dweller with limited time and carpentry skills while Bill Yanda's workshop approach to building solar greenhouses is ideal for experienced do-it-yourselfers. Steve sees a need for a solar equivalent of the storm window, an off-the-shelf item "which only requires you to insert part B into slot A and cover with part C." For those who are unwilling or unable to attempt the construction, the company will do it for about $200.

Can Be Taken Down

Another important feature of the Solar Room is its removability. Its heavy-duty redwood framing requires no foundation, an item that could cost as much as the greenhouse itself. Because the room is a temporary structure, it can usually be built without a building permit. Where the summer season is not too long or too hot, opening the doors to the structure will suffice for cooling, but elsewhere, unbuttoning the extrusion and taking the greenhouse off is an inexpensive way of getting rid of unwanted heat. Linda and Larry left theirs up last summer, although they made an attempt to provide some shade for their tomatoes and marigolds by attaching a piece of black plastic to the upper portion of the glazing. Unfortunately the plastic melted to the polyethylene, and Larry is puzzling over how to remove it.

The take-down feature makes the system eminently appropriate for trailers, it would seem, but sales to mobile homeowners have been disappointing so far. During August we saw trailers scattered around northern New Mexico with wood piles already appearing in the front yard in preparation for winter. The Kenins attribute the lack

Linda Kehoe in doorway between greenhouse and living area. The doors are left open during the day, allowing solar heat to be stored in the mass of the house.

of interest to the general belief that solar energy means collectors on the roof. There is also the fact that trailers in courts use a single meter and the costs are divided among the owners in order to qualify for an industrial rate, which is a fraction of what the ordinary homeowner pays. As a result there is little concern about the use of electricity and its cost.

Storage

The Kehoes do not necessarily typify the Solar Room owner. Larry worked for the New Mexico Department of Energy for four years, and so was exposed to conservation concepts. Solar Room purchasers are often unconcerned about heat storage, but Larry was willing to sacrifice space to nine 55-gallon barrels to help the greenhouse through subfreezing nights. Before winter sets in this year, for greater security he intends to add six more to bring the storage capacity up to more than 3 gallons per square foot of collector floor space.

Storage is often considered to be the limiting factor for a solar system. Steve Kenin says that when ERDA—now the Department of Energy—first did analyses on solar heating, it concluded that if you can't have formal storage you can't utilize solar energy. So the agency wrote off retrofitting as a national strategy. Recently the DOE has been granting funds for retrofitting commercial buildings and large apartment complexes, indicating some interest in old structures. Still, little concern for retrofitting residences has been shown.

Outraged at this lack of understanding, Kenin wrote to the government proposing a test situation to demonstrate the efficacy of a greenhouse with no storage other than the original house. He sold his proposal to the agency. Using $25,000 in federal money and $10,000 of his own, he showed what

he already knew: "You can get anywhere from 30 to 80 percent of your heating requirement from the sun without any fancy storage setup." The reason is that any house contains a lot of mass—foundation, wood paneling, even furniture. In addition, a greenhouse is often used as an air lock entrance and playroom, so that every time the kids run in and out they let warm air into the house instead of cold. Heat loss from air exchanges is reduced drastically.

Informal heat storage works best if you can allow your house to become really hot. Larry and Linda are both at work all day and the children are at school or the babysitter's. So the Kehoes just leave the door to the greenhouse open and the living area heats up to 90° F. on sunny days. By the time they come home in the evening, the temperature is in the low 80s or high 70s. Thanks to the barrels, they can enjoy working in the greenhouse after the kids are in bed. The "grab-all-the-heat-you-can-get" approach is also appropriate for older

The easily assembled 39′×7′ solar room is used for growing vegetables all winter and provides over half the Kehoes' house heat.

people who don't mind the warmth during the day and appreciate an unused furnace at night.

Steve looks forward to the day when houses will be built to retain heat, with several layers of Sheetrock on the ceiling and brick walls with exterior insulation. Soon we should be able to buy materials in our local hardware store that will provide storage on the interior of a house in an inexpensive and unobstrusive way, such as the ceiling tiles containing eutectic salts being developed at MIT.

Problems

The Solar Room, like everything else, is not without its problems. The Kehoes, for example, received a bad batch of aluminum extrusion in their kit. "Every time I turn around I have to climb on the roof and put the extrusion back together,"

The children's sandbox is on the right. At left are 55-gallon water drums that keep the greenhouse from freezing.

Larry complains. "I understand the company now has a new type and will replace ours, but I've been too busy to get around to calling them." Then there was the time the whole thing collapsed into a heap in the backyard. A nylon cord is supposed to run from one end of the structure to the other to make sure the frame stays taut, but Larry neglected to tie it on. A big wind came up and pretty soon a pin fell out, then a strut, and in a classic chain reaction the ground was soon littered with plastic and aluminum. When Larry started to put it back together, he found a few tears in the polyethylene which he patched with tape. Because the tape failed to hold well, air escapes and the plastic now sags somewhat. But at today's prices (3 cents a square foot), the entire glazing could be replaced for under $50. "Ours may not be the best example you can find of a Solar Room," says Larry, "but I guess it represents a lot of experience."

An improvement on their heat distribution method, which is in the planning stage, involves cutting holes from the greenhouse into the living areas. High and low vents would facilitate convection, allowing hot air rising off the top of the solar structure to flow up into the house and cold air to return across the floor. The continuous flow should not be rapid enough to cause a draft if the vents are sized correctly.

Small Investment

Despite the hassles, the Kehoes have not been tempted to turn in their Solar Room for the money-back guarantee. Their investment was small enough that they don't feel like complaining. The cost of the structure ranges from $6.50 per square foot of collector area for the 12-foot model to $3 for the large one the Kehoes bought. A $1,000 system can heat a thousand square-foot house on a sunny day. If the rooms were mass-produced, the price could be kept below $600, Steve Kenin claims. The Kehoes went for the big one because, although it costs almost twice as much as the smaller model, they were only talking about a difference of $500. To create the same space in glass, which would be a permanent structure and therefore subject to building codes, you would spend up to $5,000 for glass and a foundation. The payback period would balloon to as much as fifteen years.

The solar room is a pleasant place in both summer and winter.

The Kit Carson electric cooperative in the Taos area is interested in testing the performance of add-on greenhouses with informal storage because they want to cut peak load demand. If houses can be kept warm until after the dinner hour, fewer power plants will have to be built. The Solar Room company has found that if the temperature of an ordinary house is allowed to reach 85° during the day, supplementary heating may not be needed until midnight. In some locations, however, wind infiltration would limit the effectiveness of this approach.

New Mexico is a hotbed of solar activity—perhaps 20 percent of the solar homes built nationally are located there—yet the Kehoes' neighbors were nonchalant about the arrival of the innovation to their block. One indicated she didn't mind the greenhouse if it didn't become shabby looking. Most don't even realize it's there, since it is not visible from the road. The polyethylene, besides being cheap, also affords privacy in the backyard, which the ubiquitous fences were intended to provide but don't.

Many people resist the notion of polyethylene; they think they want clear glazing for views. The anticipated development of a silicon skin that is clear and long-lived should satisfy them. But they may not take advantage of views anyway because the greenhouse may well be too hot to enter and the strong sunlight may hurt their eyes. Plants and vegetables prefer the diffuse light of polyethylene since it does not create localized hot spots. In fact Linda regrets having planted so much because the greenery is growing too fast for her to keep up with it.

Solar Room is one response to the need for an easily reproducible solar-heating device that is reliable and understandable to the average American. The Kenins are also working on a "solar window" based on the same principle: two layers of inflated polyethylene that could be tacked on any existing wall to produce a little extra solar heat. Steve foresees a burgeoning of solar innovation in the next few years: "When people realize that the sun rises in the east and sets in the west, offering us hundreds of thousands of Btu a day, there are millions who are going to start trying to take advantage of that fact."

Third Time's a Charm

To him whose elastic and vigorous thought keeps pace with the sun, the day is perpetual morning.

Henry David Thoreau

Herman Allmaras is putting a third solar-heating system on his house. Not because the others have failed, but in order to test his evolving theories about heating with the sun, theories that have been changing ever since he first built in 1964. A professor of physics and math at Mesa College in Grand Junction, Colorado, he is driven by a desire to make his house 100 percent solar heated.

The ultimate goal is to cut all utility lines running into his house. Why? Fuel shortage is not the problem: western Colorado has many sources of energy—uranium, coal, petroleum. "I just don't like the idea of being tied to an outside fuel supply.

I really object when someone uses me, which happens when the utilities raise their rates without the approval of the people." Herman sees himself as part of the "Small is Beautiful" movement, which opposes the creation of ever-larger entities for the sake of a mythical super-efficiency.

Back in 1964 few people were thinking about the issue of energy independence. In pursuit of his doctorate at Colorado State University, Herman read all the articles on solar energy he could find, but there was little practical information on how to put theory into practice. He was obliged to develop his own approach to creating a workable system. Because he wanted a conventional-looking house, he decided to use his roof as the collection surface: the south slope gave him 800 square feet of collector.

Trickle Collector

A trickle system with water and rock storage was installed. Herman used recycled materials such as plate glass salvaged from broken windows. A glass dealer was paid 15¢ a square foot to cut the scraps to 10-inch widths. There were close to 200 small pieces, but their size proved to be an asset because they were easy to handle while climbing a ladder to the roof. Furthermore they have demonstrated a resistance to damage by hail and sonic booms.

A water system, Herman argued at the time, has a higher heat capacity and less dust than forced-air. In his system the rocks absorbed and retained heat from the water to keep warming the house for ten days before becoming completely exhausted. The system was purposely oversized because Herman was expecting some day to finish the attic, adding a children's playroom. Having the huge

collector has meant no worries about efficiency: "I know the collector is aging and now is probably functioning at 30 percent efficiency, but I still get lots of heat. After all, I've got 800 square feet of collector."

In order to make the heat stretch further, Herman minimized windows on the north side and used double-pane glass throughout the house. The walls are 6-inch thick sandwiches of celotex, fiberglass, and wood paneling for an R value of 20. At the time it seemed like plenty of insulation, but Herman now finds that he should have used even more. It was all he could afford as he was paying for materials as he went along. Today, he estimates, the insulation—and the house in general—would cost three times as much.

Air Collector

Modification of the trickle collector began in 1974 with the conversion to an air system making use of heat lost to the attic. A fan was installed to bring air off the collectors into ducts beginning in the attic. Phasing out the water system eliminated the

---

TECHNICAL DATA

Owner-designer-builder: Herman Allmaras, Colorado

General Features
Latitude: 39° N
Degree-days: 5,600
Insolation: 210
Heated area: 1,600 ft^2
Year of completion: House with trickle system, 1964; air system, 1975; passive system in progress
Insulation: Walls: 2" celotex and 3½" fiberglass
        Roof: 3" polyurethane foam
        Foundation: 3" urethane foam
Orientation: 15° W of S
Solar system: Originally active liquid, now active air, being converted to passive

Collection Systems
  Active Air System
  Collector: 800 ft^2 (formerly trickle)
    Angle: 57°
    Cover: Single layer of glass
    Absorber: Corrugated aluminum, nonselective coating

Storage System
Container: Steel tank surrounded by rocks in concrete bin
Material: 1,600 gal. water
Location: Basement

Distribution System
1/20-hp fan puts air into storage; air is distributed through house by natural convection.

Passive System
Collector: 400 ft^2 greenhouse
Angle: 90°
Cover: Double layer of glass
Absorber/storage: Adobe thermal wall and floor
Distribution: Natural convection

Auxiliary System
Backup: Propane space heater and fireplace
Fuel consumed: $100 in propane average

Costs
Original structure: $10,000
Original solar: $1,000
Estimated cost of passive system: $1,000

pump but added a centrifugal fan, which uses more electricity.

The air system can never suffer from the problem of corrosion. It did take some time to bring the new system up to par because the rocks had accumulated moisture from the water collector and had to be dried out. On the other hand, he lost some of the humidifying value of the water. The ducting system had to be rerouted too, after Herman learned that you have to take the air out of the same end of the storage that you put it in. In other words, you need to reverse the flow to gain the maximum amount of heat. After this change, the air system proved to be just as effective as the trickle system, and more reliable.

The biggest problem of the construction phase was selecting materials. Herman wrote to twenty companies about black paint but none would make any guarantees. Finally he found an industrial vinyl at $12 a gallon that would supposedly adhere to aluminum and resist solar radiation. He applied three coats on top of a chromate primer. After twelve years the absorber has passed its expected lifetime and the paint is discolored from the old water system, but still is not peeling. The vinyl support strips under the edges of the glass have shrunk, however, creating air leaks. By repainting the absorber and replacing the molding, he

Back of air collector is insulated with Fiberglas.

could perhaps restore the collector to its original 60 percent efficiency, but since Herman has gone on to his new project, refurbishing the old rooftop system will have to wait.

If the trickle collector did achieve 60 percent efficiency, it was doing quite respectably since no flat-plate collectors attain more than 60 to 80 percent. This refers not to the percentage of the household heat load supplied, but to the efficiency with which a collector absorbs radiation from the sun. A certain percentage is lost from reflection off the cover. Additional radiation is lost by conduction through the insulation at the back of the collector and convective air movements within the airspace between the cover and the absorber. The type of absorber coating, whether selective or nonselective, further affects the amount of radiation absorbed. So does the quantity of edge trim, which determines the amount of shading on the collector surface.

The vinyl stripping supporting the ton of glass covering the collector has shrunk, leaving air spaces that reduce collector efficiency. With 800 square feet of collector surface, Herman still gets plenty of heat.

Switch to Passive

Herman has decided to convert to passive solar heating. Attending the Second National Passive Solar Conference in Philadelphia, he learned there's no reason why a solar house should cost any more than an ordinary one. Unfortunately Herman can't start from scratch but has to work with what he's got. Retrofitting is harder, psychologically as well as intellectually, than building afresh. By installing a thermal storage wall, for example, he will lose two windows looking out on the Umcompahgre Mountains. "When you've already got something you like, the changes come hard," he says. The visual stimulation he misses will be replaced by another kind of satisfaction, that born of reducing dependency.

Herman began the passive project by adding insulation in an attempt to cut his heat requirement from 21,980 to 7,000 Btu per degree-day, which would allow him to junk his gas furnace.

A greenhouse will trap heat to be stored in an adobe floor and thermal storage wall. Constructed in front of the existing south face, the wall will provide 90 percent of the storage capacity, and the floor the rest, he estimates. The system will be tested for a year before other improvements are made: "You can get numbers from a book that give you ball-park figures, but you have to take into consideration the peculiarities of your specific environment: wind, shading, etc." Since those figures can be way off, you should test your system before you start changing it.

Using inexpensive materials such as adobe will allow Herman to keep the cost of the solar greenhouse under $1,000. By shopping wisely and accepting odd widths, he secured 1,500 square feet of plate glass for a mere $210, saving nearly $1,200.

Since his average yearly propane bill of $100 will be eliminated, the cost of the greenhouse will be amortized in less than ten years. Trying instead to increase the overall efficiency of the active system might cost as much as the original anyway. That cost has already been paid back. Herman believes that within eight years his fuel savings will equal the initial cost of the entire house.

The 18-inch adobe wall is being constructed by the "slump forming" method which involves pouring a mud mixture into forms erected on the wall site. In the course of conducting tests on adobe samples, Herman discovered that the clay on his property requires the addition of sand. By also adding asphalt he came up with an adobe block that can't be scratched with a fingernail. The movable form method is fast but he has to watch drying rates and avoid adding too much mud at once, or the wall will sag and crack.

Airspace Between Walls

Between the thermal wall and the house an 18-inch airspace will remain. Air will be drawn by natural convection up through vents he will cut in the house walls. It will flow between the joists, and

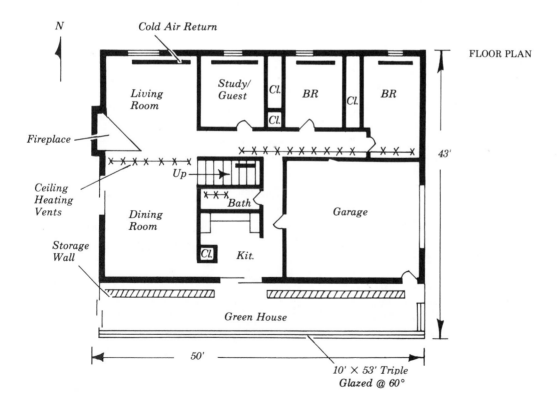

FLOOR PLAN

181

be distributed down through registers in the various rooms. When cooled it will fall and return to the greenhouse via openings in the bottom of the wall. By using a substantial adobe storage wall Herman hopes to obtain an 18-hour time lag in the transfer of heat from the greenhouse to the house. In this way he will be able to provide the energy for twenty-four hours of heat loss from the six hours of sunlight available in winter.

Although the clay in Herman's yard may be particularly suitable, adobe can be used almost anywhere if properly stabilized. Adobe is traditionally made with straw as the ingredient responsible for rapid drying, but because organic matter deteriorates over time, it may be omitted and asphalt or cement added to produce a strong, waterproof wall. Where the wall is neither load-bearing nor exposed to the elements, great strength and water-resistance are not required. In some localities building codes do not take into account new developments in adobe manufacture, so it is best to inquire before building.

Developing a solar system has been Herman's dream since he attended an Engineer's Day at the University of Wisconsin and saw an exhibit of solar cookers designed by Farrington Daniels for use in India. He remembers Daniels's report that the Hindus failed to accept the devices because they were accustomed to cooking in the time-honored traditional way. Herman concluded that "the greatest problem of all is habit," and he determined not to let conditioning stand in the way of developing self-reliance. It never bothered him that people called his house a "solar barn" and made jokes about throwing stones at the glass. His wife (they are now divorced) suffered from all the notoriety associated with owning the first solar home in the area. These days, with thousands of solar-heated buildings, owners are less likely to be considered weird although in some cases they may still be deluged by visitors. Herman warns that the attitude of energy conservation affects all aspects of one's life-style and must be shared by all members of the family.

Forging ahead with his solar program, Herman looks forward to investigating additional conservation measures. His first step will be to remodel the living room fireplace to make it an efficient backup while he works out any quirks in the new passive setup. Either a biogas system converting sewage sludge to methane by means of anaerobic digestion or a wind generator—or a combination of the two—may enable him to cut those power lines. Perhaps he could make alcohol from the grain grown on his twenty acres and use it to power lights, as well as for heat and engine fuel. "If we could make our own fuel for our cars, we could also unchain ourselves from the multinational corporations that control the world petroleum market." For voicing such opinions Herman Allmaras has been labeled an isolationist and an anarchist. So far as he's concerned, however, he's just a classic conservative trying to preserve our natural resources and his own integrity.

Adobe wall when it was being constructed by the slump-forming method. The adobe must be allowed to dry slowly or sagging and cracking result. Since this is not a load-bearing wall, the cracking is not of great importance.

Closet Architecture

Our grandchildren will live in a world without oil.
B. J. Brinkworth

During the height of the 1973 energy crunch, Kay and Dick Johnson went to a party. A guest, learning that the Johnsons were about to build a new home, casually suggested they ought to have solar heat. Thus, according to Kay, began a memorable venture into sun dwelling.

Kay, an artist and illustrator, is working on a children's book on solar energy. Dick, a professor of English at a small college in central Massachu-

setts, is interested in architecture as a hobby. He describes himself as a "closet architect," meaning not that he's a secret designer who has gone public, but that he enjoys designing closets containing spaces for everything. He showed us an example crammed full of skis, tennis rackets, ice skates, and croquet mallets, equipment for each sport in its own cubby hole.

In 1973 the Johnsons, a busy and likable couple,

and their two small children were living in a big Victorian house with large south windows. Basking in that natural heat started Dick thinking about the sun's power. So he designed a three-story saltbox with the idea of integrating collectors with the roofline. The house was constructed in 1974 and the solar panels added in 1976.

Machine on the Roof

The Johnson home is lovely, but a classic example of thinking of solar heat in terms of a machine on the roof rather than overall design. It must be remembered, however, that they planned their home back before people were becoming aware of energy-conscious design. High ceilings mean that the house contains a tremendous volume that has to be heated, in addition to nearly 3,000 square feet of living space. Dick now recognizes that cheap, abundant fuel led to the illusion that we can build houses of any size, but solar heat suggests a multifaceted approach, including keeping your house compact. "For every square foot you add," he says, "that's another square foot you have to heat."

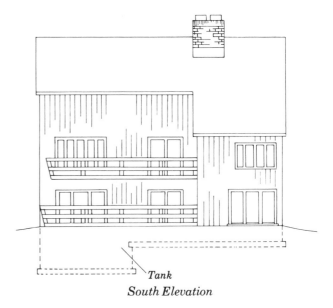

Tank
South Elevation

Inefficiencies

Kay and Dick's home has other energy inefficiencies. The Johnsons told their contractor to put in maximum insulation, not having any idea at the time how much that meant. They took the contractor's word for it, and he insulated the house the

---

TECHNICAL DATA

Owner-designers: Kay and Richard Johnson, Massachusetts
Solar consultant: Total Environmental Action

General Features
Latitude: 42° N
Degree-days: 7,300
Insolation: 125
Heated area: 2,850 ft^2
Year of completion: 1976
Insulation: Walls: 3½" fiberglass
           Roof: 3½" fiberglass
           Foundation: 6" fiberglass
Orientation: S
Solar system: Active liquid drain-down

Collection System
Collector: 610 ft^2 gross, 535 ft^2 net
  Manufacturer: Sunworks, Inc., New Haven, CT
  Angle: 54°
  Cover: Single layer 3/16" tempered glass
  Absorber: Copper sheet with copper tubes soldered 5"-6" on centers, selective coating, 2½" fiberglass
  Pump: ¾ hp

Storage System
Container: Concrete septic tank with waterproof liner
Material: 1,600 gal. water
Location: N side of basement
Insulation: 5" polystyrene on inside

Distribution System
Hydronic system: a second pump circulates heated water through 300' of baseboard radiator.

Auxiliary System
Backup: 80,000-Btu oil boiler and 2 double fireplaces
Fuel consumed: 500 gal. oil and 1 cord at $50/cord

Domestic Hot Water
Water from collector is circulated through heat exchanger in a water tank before going to storage tank. Boosted by oil boiler.

Costs
House: $65,000
Solar: $20,000+
Operating costs: Less than $20 a year

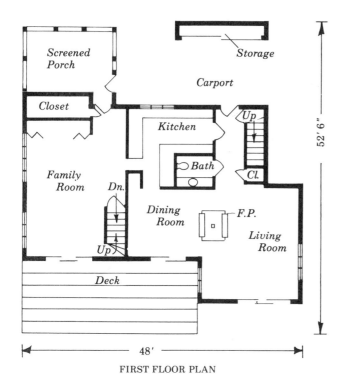

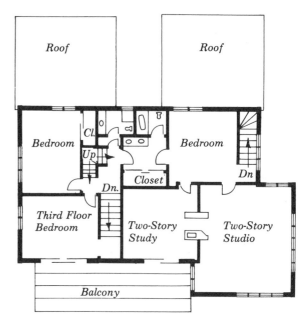

FIRST FLOOR PLAN   SECOND FLOOR PLAN

same as he would a standard dwelling without solar heat: a single layer of fiberglass on walls and ceiling.

One thing Dick muffed on, he admits, was the design of the fireplaces. The fireboxes should be shallower. They're raised off the floor too, so the heat rises and your feet get cold. The brick and concrete block chimneys, however, are inside the house and add thermal mass for heat storage.

Although it has storm windows, we thought the house employs too much glazing to be truly energy efficient. In addition, a roof formed by a second-story porch cuts off passive heat gain from south-facing sliders on the first floor. At the time they decided on the deck, the Johnsons weren't thinking "heat," they were thinking "porch."

Greater Use of Passive

Among the improvements they would make if they were doing it again would be to plan a greater use of passive heat. Passive ideas are attractive because they are often simple enough for the home-owner to dispense with hired labor. "What we've got," Dick says, "is too technically complicated to do ourselves or for us even to know whether anybody is pulling our leg."

Solar heating, the Johnsons say, was presented

to them as something they could do very simply and have it pay for itself in five or six years. In Kay's phrase, they went into it "starry-eyed." She describes their original solar consultant, the friend of a cousin, as a "brainy dilettante," a MIT dropout who had worked on nuclear submarines and designed a cement boat that sank in the Connecticut River. Initially he made a good impression on the Johnsons and convinced them that he could build a collector for $6,000. After a year and a half of their time and money, however, he called up and quit—leaving Dick and Kay stranded with a half-finished heating system.

At this point they turned to a New Hampshire architectural firm that specializes in solar heating, Total Environmental Action (TEA), who redesigned the system and ordered panels from Sunworks, Inc. A storage tank and three-zone distribution system with a thermostat on each floor had already been installed. The zone heating, the Johnsons have found, would be better organized if the bedrooms had been grouped together for one zone, and the rooms like the studio and family room that are used in the daytime were on another thermostat.

Kay said both TEA and Sunworks were better at giving theoretical advice than practical help. "Sunworks," she says, "actually has a policy that

185

they won't help you install the panels. I don't know if it's because they don't want to be liable or what." Her teeth really grated at the idea that Sunworks was willing to sell them the panels and take their money, but that's all.

Kay had to take over finding workmen to do the installation, although TEA interviewed the candidates she found. This turned out to be time-consuming: finding names, following up leads, talking to people. Both Kay and Dick are angry at TEA for charging a whole day's consulting fees for sending a man down from New Hampshire who they say stood around watching and taking pictures. The contract with TEA called for the firm to oversee installation of the panels, but Kay felt like mailing them a bill for consulting.

Kay Johnson in the living room. If she were building again, she would use the Arkansas Construction method —build with 2×6s, which can be spaced farther apart so that less lumber is needed and more room is provided for insulation.

The Water Tank

During the shakedown period the storage tank developed a problem. The automatic refilling control malfunctioned, Dick says, allowing unlimited supplies of water into the tank. "I heard a gentle slushing sound and went down and saw the cellar flooding."

We noticed the basement floor has puddles, but these are not due to flooding. Steam builds up inside the concrete tank and seeps out around the sides. The Johnsons wish they had used a steel tank and installed it below floor level so that if it leaked, the water would go into the ground.

Backup Too Small

During the first two winters before the solar system was installed, the electric backup system proved to be too small since it was working full-time but not heating the house. During those months Kay discovered the value of a communal fireplace in family life and also learned to care about the weather. She would listen intently to the forecast because she knew if the temperature went to zero, they would be unable to get the house above 58° F. Fuel bills were so high that they replaced the electric auxiliary with an oil furnace.

What with installing a second backup system and having the solar setup designed twice, costs

Dick's study with double fireplace.

skyrocketed from the initial estimate of $6,000 to somewhere over $20,000. With few errors the system could be duplicated for $12,000 to $14,000. Luckily for the Johnsons, a timely inheritance saved them from sinking heavily into debt.

It was also fortunate that the house itself was a bargain at $45,000. Dick attributes this to the contractor's enthusiasm for building a solar house and consequent leniency on charges. They saved on service calls because the heating subcontractor was willing to iron out the bugs—small things like faulty valves—for free.

Substantial Savings

Despite their recitation of troubles, their fuel bills are lower than they would be without solar heating. By keeping the house at 65° during the day and 62° at night, Kay and Dick hold their seasonal oil bill to about $340 with another $50 for a cord of wood. Considering the size of the house, the high ceilings, the number of windows, the poor insulation, the inefficient fireplaces, and the Massachusetts climate, Dick guesses that without solar heating it would cost between $900 and $1,400 a year. He bases this on conversations with friends about their fuel bills. Savings are at least $500 and maybe as high as $1,000. So the system is definitely saving money, although not when compared to paying off the original investment.

Having spent $20,000 Dick is acutely conscious of additional expenses. He points out that if you were to fully insure a solar investment of $10,000 or $15,000, you would wipe out a significant percentage of your fuel savings right at the start. He is in favor of the government providing assistance through tax benefits and direct low-mortgage loans for installing solar collectors. We pointed out that keeping your initial investment down would solve the insurance problem. Despite his experiences, Dick would rather not see the government get into setting equipment standards, for fear of overregulation.

Kay was the only solar homeowner we interviewed who expressed any uncertainty about solar heat if she were building another home. It's not that she doubts that it works, or questions whether

Solar storage tank and pump in basement.

we're running out of oil, but she would like to see costs come down before committing herself a second time. She warns that "anyone as ignorant as we were" should not get into it and advises people to remember that it's innovative, experimental, full of hassles. Check out your self-appointed "experts." Kay claims that if she had a voodoo doll of their first consultant, she'd stick pins in it. She also believes that solar consultants should find out exactly how much their clients know and not take for granted that something is understood.

Dick, on the other hand, harbors no reservations about building a solar home again. No one, after all, is ever completely satisfied with his house. At the same time, he acknowledges that their friends may have been put off by the problems he and Kay have encountered and the large investment they've made. In going to their friends for solace, the Johnsons have perhaps created a localized solar backlash, but their candor can help the rest of us identify the pitfalls.

Doing the Improbable

Criticism is easy, art is difficult.
Destouches

Dorothy Maxwell's house presented a real challenge for solar retrofitting. A tract house in a Boulder subdivision, it was built with practically no insulation. Also it is oriented the wrong way. Although the building is L-shaped, neither leg faces due south—a common enough problem with retrofits. According to Paul Shippee, president of Colorado Sunworks, which undertook the retrofitting task: "We showed that it is possible to solve these problems with a degree of success, but because of the difficulty the system is not as efficient as some."

Silver-haired Dorothy has long been interested in energy conservation. We met her returning from a video-taping session at a local television station. She had been invited to talk about her energy-saving tricycle, which she had recently equipped with a small battery-operated motor. Three wheels provide stability and allow her to carry packages and groceries. The little motor—which resembles the starter motor on a car, only instead of turning a flywheel it has an arm that helps the tires turn—permits her to go longer distances, saving gas and wear and tear on her car.

Since she lives in a high-sunshine area, solarizing her home seemed like another good way to put her beliefs into action. After her husband's death, Dorothy found herself with a sum of money that wasn't drawing much interest. She hoped to save enough through solar heating to compensate for interest lost in paying for the system. The price of natural gas remains so low, however, that the savings have not proved as great as she expected.

Maxwell house. Collector panels are whitewashed in summer to prevent overheating.

188

Still, in the first nine months after the installation was completed, she used 60 percent less gas than the preceding year even though the winter was colder. She has a good feeling about the conservation aspect of the venture, and is not too disappointed about the financial. Now that federal price controls on gas are being phased out, her savings may soar.

Savings would have been greater if passive heating had been possible. Dorothy had no south-facing walls in which windows could be cut without sacrificing privacy. Adding on a greenhouse would have meant cutting an awkward entrance into a bedroom and chopping down young trees in the front yard. An active system was thus called for, but how to put flat-plate collectors on a roof that slants the wrong way?

Curve of Collectors

The ingenious solution took the form of twelve collector panels arranged in a 70-foot curving array to catch the sun as it moves from east to west in the course of a day. Besides being functional, the upsweeping arc formed by the collectors lifts the

house out of the ordinary. The panels are supported by a "speed rail" system: steel pipe is fitted into sockets and locked in place with set screws in order to attain the angles required to support the collectors in the curved configuration.

Insulating the back of the collector panels is a layer of ceramic fiber protecting 2 inches of rigid urethane foam. Ceramic fiber was used because at the time Paul was unable to find a fiberglass with resin that could resist the collector's 200° to 250° daily temperatures. Urethane foam also condenses into a goo on exposed surfaces and emits cyanide gas at high temperatures, but the ceramic fiber, commonly used to insulate boilers, protects it.

Water circulated through the solar panels is drained into a small storage tank in the garage. A vinyl bag made for tanks is cradled in three layers of urethane for insulation and puncture-prevention. Then comes plywood with 2×4 braces nailed horizontally for reinforcement and spaced closer together near the floor to support the weight of the water. Fiberglass and Sheetrock complete the Dagwood sandwich.

The solar system was hooked into the house

TECHNICAL DATA

Owner: Dorothy Maxwell, Colorado
Solar Consultant: Paul Shippee

General Features
Latitude: 39° N
Degree-days: 5,400
Insolation: 185
Heated area: 1,200 ft^2
Year of completion: 1977 (retrofit)
Insulation: Walls: 3½" cellulose fill
　　　　　　Roof: 10" fiberglass
　　　　　　Foundation: 3½" fiberglass
Orientation: SW
Solar system: Active liquid drain-down

Collection System
Collectors: 216 ft^2 in 70° arc
　Manufacturer: Site-fabricated by Colorado Sunworks, Inc., Boulder, CO
　Angle: 58°
　Cover: Double layer of window glass
　Absorber: Copper pipe on copper backing, nonselective coating, 1" ceramic fiber and 2" urethane foam
　Pump: 1/12 hp

Storage System
　Container: Wood frame tank lined with 50-mil plastic
　Material: 400 gal. water
　Location: Garage
　Insulation: 6" foil-clad rigid urethane and 3½" fiberglass (sides) and 6" urethane foam (top)

Distribution System
　Forced air: Another pump leads from tank to heat exchanger on supply side of existing furnace; furnace controls actuate blower for solar-heating mode.

Auxiliary System
　Backup: Natural gas forced-air furnace and fireplace
　Fuel consumed: 6,000 ft^3 of gas in December, 1978

Domestic Hot Water
　Coil immersed in solar tank preheats water for 30-gal. gas heater.

Costs
　Solar system: $8,000
　Insulation: $1,000
　Operating: $6 a month for electricity for the blower.

furnace, with controls modified to permit utilization of the blower in the solar-heating mode. When heat is needed, the first stage of a two-stage thermostat turns on the furnace fan and simultaneously activates the pump from the solar tanks. If the house temperature continues to fall, the second stage clicks off the solar pump and ignites the furnace flame. When the house warms up, the second stage disengages and the system reverts to solar.

Admitting that the system is too complicated, Paul says, "If the solar heat is marginal, the fan will run all day long." Constant cycling is one liability of working with low temperatures. Dorothy pays a penalty in terms of an electric bill which increased 50 percent this year, due to the $6 monthly cost of running the furnace fan. If hot water baseboard or radiant heating had been installed in the house originally, it would have adapted more easily to the liquid solar system.

Warm At Last

What's important to Dorothy, though, is that finally she is warm. She has worn sweaters around the house for years, ever since she learned that we were running out of natural gas. Previously only the dining room, located over the furnace, was warm enough for sedentary activities such as weaving. Today she still keeps the thermostat set at 65° F. during the day, turning it up to 67 in the evenings, and down to 58 at bedtime—if it is any lower the fan has to start up in the morning. The thermostat has a red light indicating when the gas is on, so Dorothy can tell at a glance whether she is being heated by the sun or a fossil fuel. She enjoys manipulating the thermostat: "I'm not the kind of person who just wants to flip a switch and never have to lift a finger again."

In order to bring the size of the solar system down to a manageable level, allowing the collector to fit on the roof, the house was insulated as thoroughly as possible. Dorothy was amazed to find how skimpily it had been built. Despite being located in the foothills of the Rockies, it had no insulation in the walls and only 4 inches of fiberglass in the attic. The greatest insult: no dry wall under the paneling in the living room. This situation, not an unusual one, was discovered when the insulation subcontractors began blowing cellulose into the walls and a piece of paneling popped out, spilling a cellulose snow onto the rug. Dorothy suspects that the only reason the house could be heated before was the lack of windows on the north side.

By the time the insulation work was finished, Dorothy could report a noticeable difference. The

*Dorothy Maxwell
in her living room.*

190

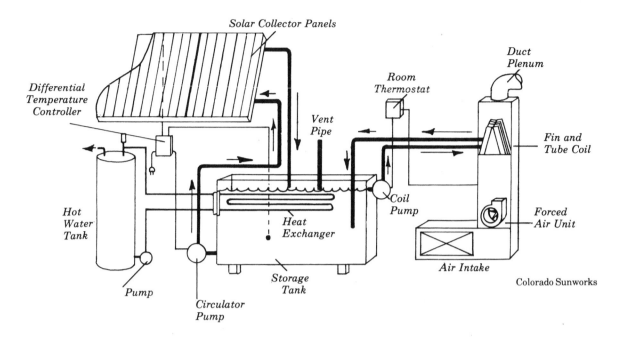

Solar heating system.

house no longer responded quickly to changes in the outside temperature. The next winter fuel consumption was cut in half. The solar system is responsible for saving another third—but the insulation was only one tenth the cost of the solar. According to Paul's calculations, the payback period is sixteen times better for the insulation. Hearing this, we questioned Dorothy's going to solar at all. Paul's reply was that you can't just look at the initial installation expense. Savings must be judged over the long term: "Putting in a solar-heating system is equivalent to buying almost all your fuel for the next twenty years at once."

Freely admitting that the Maxwell system is not the most economical even in the long run, Paul is still justifiably proud of having devised an active system which at least partially meets the need in this situation.

For new construction he is turning to completely passive systems: "I think flat-plate collectors are going to be obsolete for new homes by 1980," he says. "At first there was a great euphoria about solar heating. Everyone jumped into active systems without knowing much about them. Now they've been in operation for a couple of winters and we're finding they have a lot of problems." Not only are there just too many things that can go wrong, but also active systems cost too much for

Front entrance with detail of collectors which are supported by a "speed rail" system of pipes and sockets.

191

The solar heating system hooks into the original forced air furnace. On the left is a 20-gallon gas water heater that stores domestic hot water and boosts its temperature if necessary.

what you get. You pay for the collectors year-round even though during half the year they give you only domestic hot water. In fact you have to run the pump to keep them cooled. If you don't, they can heat up to 300° or 400°, Paul warns from bitter experience, and materials disintegrate rapidly under those conditions.

Dorothy Maxwell, for example, gets ten times more heat than she needs in the summer. She could sell hot water to her neighbors if she were the entrepreneurial type. In order to protect the tank's vinyl liner, the collectors shut off and the water drains down when the storage reaches 160° F. When the pump comes back on, it sends water into the bottom of the copper pipes, which have been baking in the sun. The water begins to boil, flashes to steam, and starts pouring out vents at the top. It's quite a sight: sort of a miniature roof-top geyser. The water gradually cools the copper plates and things go on as before. This pheno-menon may occur several times on a sunny day if the temperature in the tank is high. In case, like us, you were wondering how the glass reacts to the steam, you can relax. Because the glazing is hot when the steam emerges, there is no danger of breakage.

Customarily the way to avoid overheating is to place the collectors at a 90° angle or to vent the excess heat, either way reducing efficiency. Mrs. Maxwell found a simple answer in a whitewash made by American Clay Co. of Denver. The coat-ing will not wash off until after the first freeze of the season. For a nominal fee she hired a neigh-borhood boy to paint the collectors. Since the whitewashing in August of this year, she has no-ticed that the pump runs during the day when it wouldn't have before, indicating that the collec-tors are cooler. She also tested the whitewash's removability and found that even after freezing it won't wash off with just a rinsing, but will have to be sponged. Another job to keep the local Tom Sawyers occupied: an unforeseen benefit of solar heating.

Of Tin Cans and Old Tires

The path of civilization is paved with tin cans.

Elbert Hubbard

It was a bright August day and we were driving back from Kit Carson National Forest to Taos, New Mexico, after a camping expedition. Suddenly as we rounded a curve we spied on a canyon side what appeared to be a thousand beer cans glistening in the sun behind a stand of juniper trees. Sure enough, as we approached we discovered a house built almost entirely of aluminum cans, with a round addition made of automobile tires going up alongside.

This was the first solar can home in the country, owner Tim Curtis told us. There have been other

Curtis house. A barbecue smoker dominates the 800-square foot deck.

can houses, but because the containers were empty, the rooms got cold and stayed cold. "Of the many means of storing heat, just why did you pick cans?" we inquired. Tim had come from Texas to Taos and had found a job as an accountant with Mike Reynolds, maverick solar architect. Reynolds had already designed and built several can houses, including his own, and had been involved in other innovative projects with bottles and scrap materials through his firm, World Energy and Materials. One thing led to another and Tim soon found himself with trowel in hand mortaring cans. It was an activity he'd never expected to engage in, but then changes were, in solar language, coming hot and heavy, including a subsequent shift in professions. Through keeping books for a gallery, he became acquainted with the art business and soon emerged as a "talent agent for artists" and opened his own gallery in town. Tim's artistic tastes are clearly evident in the eye-catching house.

Building with cans is, Tim discovered, easy work even for the novice: "Any fool can plop a can down and mortar it in." Although it goes slowly, you can start at one corner and go all the way around the house doing three or four courses in a day, so actually you can make quite a bit of progress by the time the sun goes down. Laying up the container-filled walls took about a month while the entire house was completed in three. Finishing touches are still being added.

Building With Cans

Designed by Reynolds, who also supervised the construction, the house is built with post and beam techniques, cans filling in between the posts. The exterior walls are two cans deep, with the cans laid horizontally, and with 2 inches of Styrofoam in the middle of the wall. The cans on the outside contain only air, for insulation, while the interior ones, filled with water, provide thermal storage. A

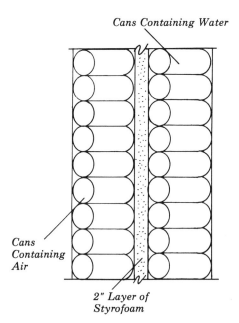

Schematic of exterior wall.

Phoenix brewery filled the cans and sealed them. Reynolds eventually bought his own sealer, but capping turned out to be uneconomical in terms of time. Lining the cans up perfectly is unnecessary; besides, some day Tim may stucco the interior of the house to give it the adobe look popular in New Mexico.

Other parts of the construction were harder. The

TECHNICAL DATA

Owner: Tim Curtis, New Mexico
Designer: Mike Reynolds
Builder: Tim Curtis, Michael Tennyson

General Features

Latitude: 36° N
Degree-days: 8,000
Insolation: 250
Heated area: 1,500 ft^2
Year of completion: 1976
Insulation: Walls: 2" Styrofoam between aluminum cans
           Roof: 9" fiberglass
           Floor: 1" cellulose board
Orientation: S
Solar system: Passive direct-gain

Collection System

Collector: 150 ft^2 of double-glazed windows
   450 ft^3 greenhouse glazed with two layers acrylic

Storage System

Containers: Aluminum cans mortared into walls
Material: Water
Location: All walls of main structure and under floor

Distribution System

Heat stored in walls radiates into house; high and low vents in greenhouse wall permit circulation of warm air into house by natural convection.

Auxiliary System

Backup: Electric cables under floor and fireplace
Fuel consumed: $40 in January at 4¢/kwh average

Costs

House: $30,000
Solar: Negligible

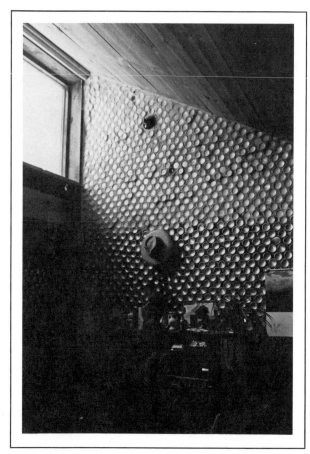

A clerestory in the bedroom admits light and heat.

dows gathers additional solar radiation, with vents admitting warm air to the living area.

The kind of heat provided by thermal storage walls is subtle. There are no cold spots except for the bathroom, which is more isolated than the other rooms and located on the northwest side of the house.

Electric Backup

An intriguing backup system beneath the living room floor consists of a layer of cellulose board on top of bare ground, followed by Sheetrock, then water-filled cans and plastic bottles. Next, a layer of aluminum foil reflects heat from electric cables lying on it and held in place by a coat of plaster. Placed so as not to interfere with the cables, 2×10 framing supports fir flooring.

The all-electric house runs up bills of $65 a month during January; of that about $40 represents the auxiliary heating load. It gets cold in the Taos ski area: sub-zero temperatures may be experienced for a month at a time, and sometimes it plunges to 30° below. An ordinary all-electric house is said to cost $900 per season, so in this case solar heating cuts the bill by at least three-fourths. A free-standing fireplace in the center of the house is made from an old boiler. The visual effect is striking, but, because of a poor draft, the fireplace is not often used as a source of heat.

Behind the fireplace, curving stairs lead up to an 800-square-foot open sun deck. Surprisingly, the first thing we encountered when we arrived on top was what appeared to be a wood stove. We puzzled silently for a moment but were finally forced to inquire about its possible uses. It turned out to be, not a stove, but a smoker. Tim graciously explained. "You take a big brisket and smoke it for sixteen hours—start in the evening and cook it all night and the next day—by evening you have some real barbecue." Barbecue originated in central Texas among the German immigrants. "They didn't use sauce—that's a perversion—they just smoked it for a long time to get that flavor."

From the edge of the deck we peered down at the greenhouse, which is only 4 feet wide because the hill drops off abruptly in front of it. The narrow width offers little room for growing plants (although the ones in the planting beds next to the

back of the house is molded to the canyon side. Although a backhoe did some of the work, the wall was dug mostly by hand. Next, the builders pushed wheelbarrows of dirt up the slope to form a terrace. Then they hauled concrete for masonry work. It wasn't an effort anyone would care to repeat.

Being partly dug into the mountain, the house has a series of levels, each 2 or 3 feet higher than the next. On the lowest level, facing south, is a 30×15-foot living room. Up a few steps is a landing. Off it are the den (with a sleeping loft), kitchen, utility room, and bath. Slightly higher is the master bedroom. Each level has its aluminum polka-dot walls.

Passive Heating

Most of the direct solar gain is achieved through double-glazed windows in the living room. Clerestories also admit light into the den and bedroom. A long, low greenhouse under the living room win-

*Builder Michael Tennyson
in living room.*

wall are doing well) but it works fine for solar heating. A double skin of fiberglass-reinforced acrylic retards heat loss at night, while the can wall provides thermal storage. For a little added flavor a bottle wall forms the east end of the greenhouse, partly framing the main entranceway. The many-colored bottles are a miscellaneous collection apparently accumulated through many parties, laid up with their necks sticking out.

Bottles are used again as windows in the house, breaking the sea of aluminum. In this application the necks were broken off, and the bottles were mortared neck to neck through holes cut in the bead board insulation. "The artistic ramifications of this technique are unlimited," according to Michael Tennyson, one of the builders and a current occupant of the house. The cans too can be used to form unusual shapes, as in the sculptured divider wall in the bathroom.

Building With Tires

Unfortunately it is becoming harder and harder to get aluminum cans. These days recycling programs grab them up, and you'd have to drink a lot of beer to build a house. The final blow for New Mexico can builders came when the Phoenix brewery stopped sealing them. Tim switched to old rubber tires for his addition, begun in June 1978. Building with tires is easier and faster than with cans although it's heavier work. Mike Reynolds originated a devil's food doughnut construction method. His technique is simply to pile the tires up in a staggered fashion, filling each with earth. A hosing with water packs in the dirt. Corners, spaces around doors, and any sections requiring extra support are filled with cement block or concrete. To create a more even surface and provide insulation, bottles are stuck in the spaces between the tires.

At the top of the wall, a 2"×12" plate is anchored with bolts embedded in concrete poured in the last layer of tires. The whole wall is covered with chicken wire and plastered, first with adobe, and then with stucco for greater permanency in the high altitude. One reason for using the tires is that at the moment cement is in short supply, and when you can get it the price is exorbitant. Clay for the adobe, however, has to be hauled in because mountain dirt is not suitable.

No foundation is required for the tire structure, says the architect, because the 2-foot thick walls are too wide to heave up from frost. The form of the building is not accidental: a curved configuration provides structural strength. On the north side the addition will be bermed with earth up to about 5 feet. On the south, a clerestory above the tires will admit sunlight and warmth, which will be stored in the mass of the walls. The only difficult task is attaching the addition to the house through a "portal." Fitting a partial gable roof into a circle is a design problem builder Michael Tennyson is still working on.

"This house doesn't represent a maximal solar situation," Tennyson says. "We do have to use some auxiliary heating." In part the lack of efficiency is due to the site; too many trees shade the house on the south, and a bluff across the road cuts off some sun in the morning and again in evening. It is not hard to keep the place at 70°, though, even when outside it's 30° below. We noticed that it stayed quite cool in the summer because of the chimney effect created by the stairway to the deck.

A Few Construction Mistakes

Only a few genuine mistakes were made in the construction of the house. The doors are too heavy for their hinges, for example, and the salvaged wood used for the decking is curling. Hand-built out of cement with a rubber-like coating, the bathtub was installed without a trap. More insulation in the ceilings would no doubt save heat. Some people might find the kitchen counters, built to fit the 6'5" Texan, a bit high.

Fireplace made from boiler lid in center of house. Kitchen is on right, stairs at rear lead to deck.

If he could build again, what would Tim do differently? "Well, we *are* doing it again," was his reply. Next door the group is putting up a wood-frame spec house with passive solar heating. This time, instead of trying to build a split-level into the mountain by hand, they had a bulldozer dig the lower level and stacked a second story on top. The bathroom was moved to the southeast corner so that it will warm up in the early morning sun; the new owner will be able to take a shower before he or she goes to work. There will be no electric cables under the floor; for simplicity an adobe fireplace will provide auxiliary heat.

Despite the unusual construction techniques, Tim had no run-ins with the building authorities. Of course, both cans and tires have been tried before..."or maybe they just don't care what the crazy artists around here are doing," Tim ventured.

It seems ironic that recycling has already brought the days of the solar can house nearly to an end after only a brief start. Now the tire house appears in its stead. Perhaps this kind of creative thinking will inspire others to new heights. Beyond rubber tires, what?

Can house as seen from interior of circular addition.

Dream House

*I do not know whether I was then a man dreaming I was a
butterfly, or whether I am now a butterfly dreaming I am a man.*

Chuang Tzu

Not every couple who can afford to build their
dream house wants to do it themselves. For Tom
and Jane Cohen, however, the dream included actually constructing their impressive redwood
home overlooking San Francisco Bay.

Having sold their first house in Mill Valley at a
time when real estate prices were climbing rapidly, they realized a substantial profit. Then Tom, a
television director whose previous building experience was limited to putting in a closet, read Rex
Roberts's *Your Engineered House* and became convinced that he could handle post-and-beam construction as well as anyone. Trained as a botanist,
Jane was concerned that the house embody eco-

logical principles. And so they set out to build
themselves an elegant energy-efficient home.

The 1965 Roberts book provides basic design
principles and, without emphasizing solar heating, teaches elements of passive design. To the
Cohens it seemed unlikely that passive techniques
could fulfill all their needs, so they looked for help
in designing an active system.

Interactive Resources, Inc. was chosen because
the Cohens admired one of IRI's projects, a nearby
solar-heated photography studio. For assistance in
designing the house, Tom turned to a friend, E.
Burr Nash. This combination of consultants jelled
well. Not long out of architecture school, Nash had

enthusiasm for the design as well as building experience, and was available to help whenever the Cohens got stuck during construction. They worked with IRI for three weeks, and then IRI's construction affiliate, Sun Light and Power, finished the project. "They certainly did a good job," Tom says, "but I'm not sure they made much money on this one. Being neophytes too, they spent too much time and care on it."

The hillside acre purchased by the Cohens is well-suited to solar heating, so much so that when IRI's Dale Sartor came to assess it, he said, "Sure, on this site you can do it ninety-five different ways." Since the southern exposure coincides with a view of the bay, a Trombe wall was eliminated from among the candidates. A greenhouse for Jane was definitely to be part of the picture while radiant floor heating also seemed like a good bet. It

was thought to be more comfortable and less noisy than forced air. And, important for the owners of two large canines, it wouldn't blow dog hair around.

Radiant Floor Slab

A radiant floor slab is well-adapted to solar because the heat transfer fluid is identical with the collection fluid and because low temperatures are acceptable. The only problem, according to Nash, is that the radiant heating has to be turned on early enough in the morning so that the slab is warming the house by the time the owners wake up. It must be turned off well before the heat is no longer needed to prevent depletion of the stored heat.

TECHNICAL DATA

Owner-builders: Jane and Tom Cohen, California
Designer: E. Burr Nash
Solar consultant: Dale Sartor

General Features
Latitude: 37° 30' N
Degree-days: 3,100
Insolation: 125
Heated area: 2,800 ft^2 plus greenhouse
Year of completion: 1977
Insulation: Walls: 3½" fiberglass
      Roof: 2" polyurethane foam with 20 mil diathon
          coating
      Floors: 3½" fiberglass and 4" airspace
Orientation: S
Solar system: Hybrid: active liquid drain-down and passive direct-gain

Collection Systems
  Active System
Collector: 400 ft^2 (36 panels)
  Manufacturer: Site-fabricated by Interactive Resources, Inc., Point Richmond, CA
  Angle: 45°
  Cover: Single glazing of tempered glass
  Absorber: Copper tubing with copper backing, nonselective coating, 3" Technifoam
  Pump: ¾-hp suction

  Storage System
  Container: 1,500-gal septic tank
  Material: Water

Location: Buried in front yard
Insulation: 3" Technifoam (bottom and sides), 2" urethane foam (top)

Distribution System
Radiant floor heat, with copper pipes spaced 6" apart in 2½" concrete slab. Powered by 1/6-hp pump.

Passive System
Collectors: 140 ft^2 of double-glazed windows on S and E
  140 ft^2 of glazing on S and E walls of greenhouse

Storage System
Cement block retaining wall in greenhouse

Distribution System
Natural convection

Auxiliary System
Backup: Gas fired boiler and 3 fireplaces
Fuel consumed: $34 in gas, January, 1978

Domestic Hot Water
80 ft^2 of the active collector heats domestic water, which is stored in a 120-gal. galvanized tank. Boosted by 30-gal. gas heater.

Costs
House: $115,000 (materials)
Active solar: $9,000

There were other design criteria as well. Tom and Jane wished to retain the best aspects of their previous home, a 1908 redwood shingle structure in the style of California architect Bernard Maybeck. The attempt was to combine the warmth and hominess of old wooden houses with the efficiency and openness of the modern. The need to tuck the house back on the tail end of an odd-shaped piece of flat land gave rise to a two-pronged design. This shape developed some strong angles, which were emphasized in visible exterior framing and in the wedge-shaped entranceway where the two sections of the house intersect. To create that old farmhouse look a broad veranda was formed by adding an overhanging roof that supports the collector.

Sizing the Collector

Because of a lack of data, sizing the solar system was difficult. The Roberts book described how to measure the sun's positions, enabling the Cohens

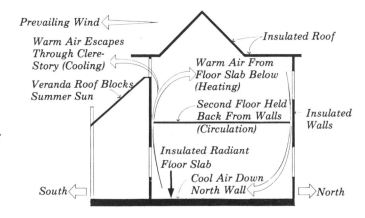

Schematic of passive heating and cooling principles.

to take readings at every equinox and solstice for a year to find out where the sun falls on their lot. Although in this way they got to know the property well, they still lacked accurate information on the local degree-days. The only data available were for Point Richmond or San Francisco, both of which are foggier and colder than Mill Valley in general, and their ridge in particular. In the end

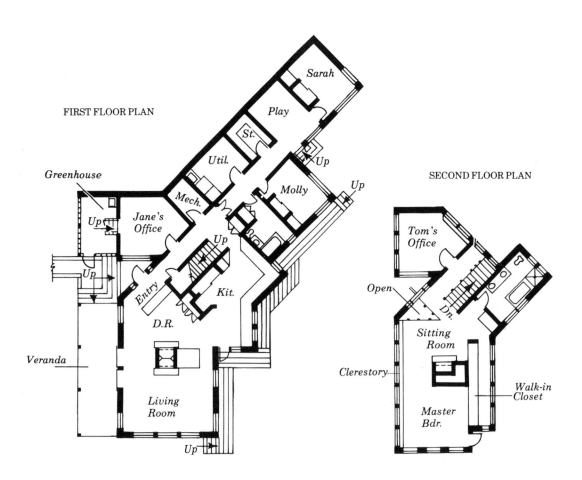

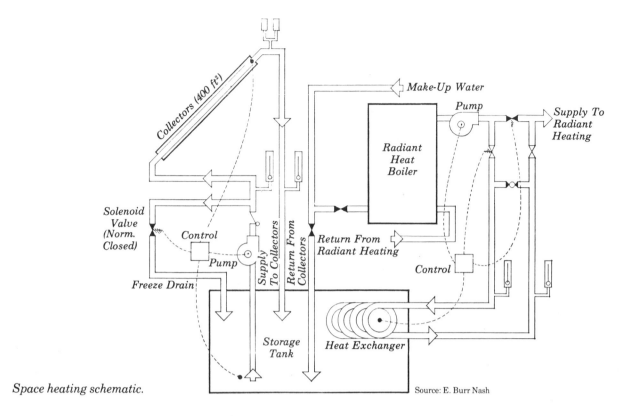

Space heating schematic.

Source: E. Burr Nash

they settled on 400 square feet of collector for space heating, a third of the downstairs floor area. The intention was that the second floor should use no direct radiant heat, as hot air would rise through floor registers. A clerestory would provide additional warmth. Since this circulation system, however, has not worked as well as predicted, the upstairs is a little cool.

The four banks of collectors are integral parts of the Cohens' roof, replacing shingles. The all-copper absorbers are covered by a layer of glass from sliding door seconds. At first the collectors leaked

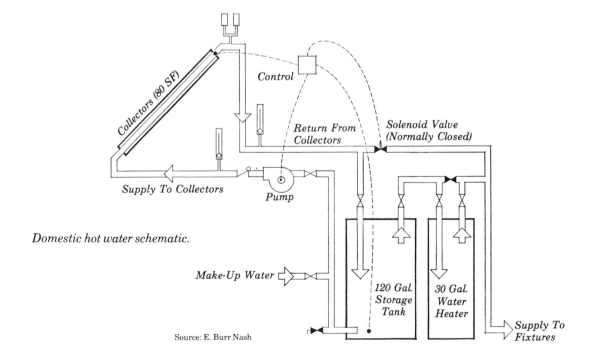

Domestic hot water schematic.

Source: E. Burr Nash

water into the house, but recaulking the joints where aluminum extrusions retain the glass eventually dried up the trickle. It is hard to make a watertight roof when there are so many joints, designer Nash says; great care must be taken with sealants, and flashing must be carefully designed. In fact, integrating water collectors into a roof seems to us inherently risky.

Storage Tank in Yard

With the solar-heated water stored in a septic tank buried in the front yard, no interior space is taken up in storage, but there are transmission losses. A septic tank was chosen because it was an inexpensive box of the right size, but the tank settled unevenly due to rain and the Cohens' forgetfulness with a lawn hose. Consequently, because the tank was unreinforced, it cracked as it was being filled. Tom twice tried to fix the cracks but finally had to call the septic tank supplier to repair them with a coal-tar sealant. Nash suggests that a reinforced concrete or fiberglass tank (one that can withstand high temperatures) would be a better choice.

Because the water had to be drained from the tank during each repair attempt, the solar-heating system got a late start on the winter season. Fortunately a boiler had been installed to boost the temperature of the water going into the radi-

Tom and Sarah inspect solar storage tank in front yard.

ant heating coils. Due to the belated startup, the Cohens derived less of their heat from the sun during their first winter than they will in the future. In January their gas bill was $34, but usage gradually dropped as the system was perfect-

The kitchen cabinets are of fir. Ceramic tile covers radiant floor slab.

Living room has east windows to catch morning light.

ed. By May the backup was shut off, and the gas bill was only $2.

Apart from the leaks, the only difficulty presented by the solar system involved balancing the collectors. With the east and west banks plumbed separately, the problem is getting them functioning at the same temperature, as they should for maximum efficiency. Theoretically this requires a one-time adjustment of the amount of water flowing through the collectors, but Tom had to tinker all winter with the mixing valves.

A radiant floor is particularly appropriate for use with a solar system because it performs well with relatively low-temperature water. It is not inexpensive, however. Nearly a thousand dollars worth of copper pipes carry the heated water to the ground-floor rooms. The house, built on land that is half cut and half fill, sits 18 inches off the ground on piers. A plywood subfloor is covered with 2½ inches of concrete, with the radiant heating pipes embedded in the concrete. A surface of beige ceramic tiles, requiring no sealing or waxing, aids in heat retention and distribution.

Located with the boiler and other hardware in the mechanical room, thermometers indicate the temperatures of the collectors and the stored water. On the sunny day in February when we stopped in, all the collectors read approximately 130° F. It had been cloudy for most of the previous month, so the water in the storage tank was only 90°, in the floor slab even cooler. The Cohens have no way of knowing exactly what temperature the floors are, but they do feel warm to the touch and heat the house satisfactorily.

Passive Heat

The amount of heat coming from the passive aspects of the house is undetermined but substantial. Sunshine poured in the windows of the living room where we sat to chat, though not far enough to strike the furniture. Clerestory windows on the second floor admit light and warmth to the master bedroom and to Tom's office. Downstairs, Jane's study is heated by natural convection via dutch doors to the greenhouse. Summer cooling is accomplished by opening second-story windows on the south and ground-floor windows on the north. The greenhouse is cooled by a foot-high vent at the top of the glass.

The key to building the house of their dreams, the Cohens say, was time. The whole project required a year and a half. Tom worked on it alone for three months, then acted as contractor and chief carpenter, bringing in a friend from the T.V. station to help. Jane spent six weeks on her hands and knees laying 1,800 square feet of floor tile.

Tom and Sarah in dining room.

Loving care was obviously taken with the finish work, for which native materials were selected. The downstairs paneling is clear-heart tongue-and-groove redwood, as is the decking of a small sun porch off the master bedroom. The kitchen cabinets are fir. All ceilings and the second floor paneling are pine. For the sake of esthetics as well as insulation, wood sash was employed on all windows and doors. An antique railroad car lavatory in the guest bath and thrown pottery sinks in the master bath add spirit to the decor.

Other energy-saving features include a super-insulated electric wall oven and bubble skylights brightening the hallway and the playroom of daughters Molly, 2, and Sara, 5, making lights unnecessary during the day. The tiled fireplaces, with built-in air recirculation systems, have separate flues inside a shared chimney. Two are back-to-back in the living and dining rooms while an upstairs fireplace serves the master bedroom. So far they have been used largely for atmosphere.

The Cohens are content with their solar dream house. Sometimes it bothers Tom that they haven't escaped dependence on electricity since the solar system is run by pumps. The original ½-hp collector pump proved to be too noisy and too small, and was replaced by a larger model. Another pump sends water through the radiant distribution system while a third powers the domestic water. The number of pumps is only a small irritation, hardly worth complaining about. As Tom puts it, "Jane and I are very pleased with the entire project and look forward to living here forever."

Solarizing a Trailer

The way out is through the door. Why will no one use this exit?

Confucius

What do you do if you want to build a solar home but you have no funds, only a piece of land in northern New Mexico and a thirty-year-old trailer? Obviously—at least it was obvious to Thor Sigstedt—you build the house around the trailer, using recycled materials that you scavenge.

Where do you start? You haul the trailer to your land and try it out here and there to determine the most agreeable building site. "The wonderful thing about a trailer is that you can move it around following the sun to find the best position for solar heating," says Thor. He chose a south-facing rise studded with junipers above an arroyo as the final resting place for his Custom Built trailer. The site had no water, no electricity, and the deed stipulated that within two years all dwellings on the property had to be permanent adobe-style structures. So Thor took up his chain saw and hammer and got to work to turn his trailer into a solar-adobe.

This owner-builder possessed skills suited to the task. The ex-stepson of Ken Kern, author of *The Owner-Built Home*, he had been building things ever since he got his first set of blocks. Now, at 25, he was a maker of fine furniture and a dedicated recycler. But he needed a shop in which to work, in addition to a place to live. His goal became to construct a woodshop on the north side of the trailer and a greenhouse-solar unit on the south, with a central living area in between. The trailer would remain as the kitchen and office.

No Time for Planning

As time was at a premium, Thor launched into the job without detailed plans. He now wishes that during the design period he had taken more care in determining his needs. On the other hand, he feels it's crucial to be realistic and flexible in your goals. Try to plan each stage carefully but be willing to make changes as you go along.

The trailer was 30×8 feet, not very big, but big enough to live in while building. Thor started by digging a foundation trench for the shop and greenhouse. Then he went to his favorite source for materials—the dump—and came back with scrap wire and metal to use for strengthening the foundation. Because of the lack of water on the site for mixing concrete, a cement truck was called in for the pour. On the south side he built a cinder block wall, reinforced with wire, and filled with concrete. Bolts were threaded in to anchor a 2×8 sill plate to the foundation.

Convective Loop

The foundation is a foot or more wide, broader than necessary, but Thor was uncertain of his next step. Perhaps he would put in an adobe wall: "All I knew was that I had the trailer and I wanted a shop and a solar greenhouse; how to do it was still unclear." Finally he realized that because the trailer is above ground he could set up a convective loop around it. How to use that principle was the next question.

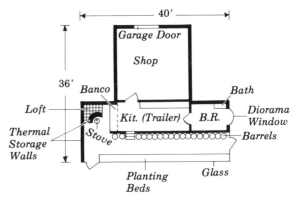

FLOOR PLAN

His solution was simple but effective. A roof was constructed a foot above the trailer so that hot air from the greenhouse rises and flows over the trailer into the shop. The cooled air is drawn back underneath the trailer to be rewarmed. Twelve manually operated vents along the bottom of the greenhouse wall provide ventilation during hot weather. Wind chimes tinkle constantly as the air circulates.

The framework for the shop was established when poles were set into the concrete foundation. The poles cost nothing but labor. Thor cut them himself in the national forest with a "standing

---

TECHNICAL DATA

Owner-designer-builders: Thor and Carolyn Sigstedt, New Mexico

General Features
Latitude: 36° N
Degree-days: 6,000
Insolation: 250
Heated area: 1,040 ft^2
Year of completion: 1977 (solar restoration)
Insulation: Walls: 6" fiberglass on N; 2" polystyrene on E and
          W (exterior)
        Roof: 6" fiberglass
        Floor: Pumice in greenhouse
Orientation: S
Solar system: Passive direct-gain

Collection System
Collector: 450 ft^2 (double glazing on greenhouse)
    Angle: 60°

Storage System
Containers: 70 5-gal. barrels
Material: Water
Location: N wall of greenhouse

Distribution System
Heat rising off top of greenhouse convects through space above trailer into shop at rear, recirculates under trailer.

Auxiliary System
Backup: 3 wood stoves
Fuel consumed: 4-5 cords

Domestic Hot Water
Wood-burning water heater (to be tied to solar preheating system)

Costs
Trailer: $200
Structure: $4,000
Solar: Negligible

Instead of adding a room onto his trailer, Thor Sigstedt built a house around it. This view shows the trailer surrounded by adobe walls with a greenhouse emerging on the south. (Courtesy T. Sigstedt)

dead" permit. One fundamental error was made, however: Thor neglected to put wood preservative on the posts before sinking them in concrete. They will eventually rot if he doesn't get around to drilling holes into the posts and filling them with preservative.

Since he had no electricity Thor was unable to use electric tools. Fortunately the post-and-beam construction method lends itself to the use of a chain saw. Any cut that is eventually covered can be made with one, as long as the measurements are correct. Barn nails and spikes, driven by a sledge hammer or the back of a wood-splitting maul, hold everything together.

Because of a lack of funds for lumber, Thor framed the greenhouse on 3-foot centers. Fortunately the glass he bought came in a standard width of 3 feet. Purchased at a cost of $40 a panel, it was the largest single expense of the whole project. But what if the glass had come in 2½-foot widths?

Thor emphasizes that luck does not always make up for lack of planning.

Use Standard Building Techniques

It is important to use standard sizes and common construction methods if you have a tight budget. Repeat the same methods and materials, he advises, because you can generally buy more of any item for less: "People get into sculptured houses, with curves everywhere, which is fine, but it's not cheap. You realize that when you start buying glass or making vents."

An example of this quantity buying approach is the solar storage system. The north wall of the greenhouse is lined with 70 5-gallon barrels full of water. They store heat in the winter and release it slowly to equalize day and nighttime temperatures. Painted brown, they are inconspicuous among the plants. During the summer when the sun is high, the insulated roof of the greenhouse prevents light from striking them, so that unwanted heat is not stored.

Finishing the House

Once the basic framework of the shop and greenhouse was in place, the house could be expanded to create more living space. It was at this point that Carolyn joined the crew and helped to design the adobe living area on the west side. "We were married in the middle of the house," says Carolyn. "I mean, in the midst of the construction, not actually inside." From then on, Thor did most of the "grunt" work while Carolyn took a job that brought in funds for materials.

The last, glorious step was cutting out the wall of the trailer to open it into the new living area. The adobe addition, insulated with rigid polystyrene on the exterior, has a floor of sand covered by flagstone. The walls and floor absorb heat from the windows, which continue as an extension of the greenhouse. A thermal wall of adobe in the center of the room stores solar heat during the day and heat from the wood stove, as well as supporting a sleeping loft above.

Wood Stove Heats Wall

"As far as I can tell, the more mass you can put into a place, the better," Thor says. "Then if you put the wood stove right by the thermal storage wall and the loft above it, you don't have to worry about being cold at night." During the coldest months the door to the shop is closed at night and the stove stoked. Sometimes a fire is built in the morning too if the sun isn't up before the Sigstedts. The structure wasn't completely sealed last winter, so the performance should be better this year. Thor is toying with the idea of reducing nighttime thermal loss through the glass by making shutters of Styrofoam attached to Masonite and hinged to the roof.

Carolyn loves the greenhouse with its view to the south. She is not a stoic: "I'm really quite

Living room, with adobe thermal storage wall supporting sleeping loft.

spoiled. I love heat. This place is comfortable, only you have to do a little work—light the stove occasionally." The temperature fluctuations are fairly large, but not so extreme as to present a tremendous problem. One of these days they'll have to break down and buy a thermometer to find out just how warm the house is getting.

Small Investment

To Thor this is not a dream house. "If I could start over again, I would do everything differently, beginning with having more money." Life is transitory, he points out. If you travel across this country you see houses falling apart along the highway— boarded up, derelict. If this house is abandoned some day, at least Thor can salvage the glass. It may be inhabited for a long time though. In Appalachia, Thor saw company houses in coal-

The greenhouse runs the full length of the building, permitting direct gain in winter.

mining towns: "They slapped up those crummy shacks and people still live in them. At least this house has some individuality because it was built by the owners."

Thor is only 26 now, and few people his age have built their own home without a mortgage hanging over their heads. He did it by scrounging, taking out small loans, and getting a little help in the form of labor from friends. You can't do it alone, according to Thor. If you want to build a truly low-cost house, call on your friends the way they used to do for barn raisings: Thor found that people love to get out in the country and join in working with natural materials to produce something tangible, especially if they are supplied with food and drink. The stucco adobe style is well-suited to this approach: you can have a party and invite neighbors out to do the plaster finish, which does not require skilled labor.

The county dump was Thor's silent friend. It offered everything from wood for the foundation forms to an old cement mixer with a lawn mower motor. Thor used to go to the dump almost daily, bringing home anything that might be useful. New federal legislation has led to the closing of open dumps: "It's a terrible crime, making dumps inaccessible to people. You could go to one and see

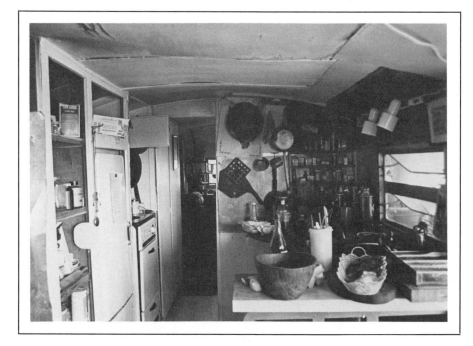

The trailer was opened to join the kitchen to the living area.

four or five families getting their clothes, furniture, appliances, everything." The dump was the major American recycling center. Now that it's gone, all those perfectly usable things will be wasted, buried under tons of dirt.

Respect for Trailers

Just because it's made of recycled materials doesn't mean the Sigstedts' house is unsightly. Some fine touches, such as an intricately carved front door and a diorama window with a desert scene, contribute to a basically handsome structure. The silver trailer will eventually disappear under wood paneling. It is almost hidden now by the barrels and plants.

When Thor meets arriving guests, he often finds them saying, "Oh, what a lovely house. You built it yourself?" Then they walk in and it's "Oh, my God, there's a trailer in here."

Many people add a room on to a trailer and get additional space, but they don't take advantage of solar radiation and the possibility of natural convection around the trailer. Environmentally, Thor says, trailers are "lousy in every respect." But you can put a second roof over one and instead of skirting the issue, take advantage of the fact that it's off the ground. Besides, a trailer can be moved around first to try out different sun angles. And as your trailer gets older and more claustrophobic, you can start tearing down its walls, slowly opening up the space. You can even build in such a way that you can pull the trailer out later if it is salable. For Thor that is not a possibility: "When I brought the trailer down here, this was its last stop. I would have had to pay somebody to haul it to the junkyard. I really had no respect for trailers then— Carolyn hated them too—but now I realize their true value."

Mirror, Mirror in the Roof

People need temperature contrasts.... If you keep them at the same temperature twenty-four hours a day, you'll turn them into earthworms. They'll become like larvae, they'll lose all their natural defenses.

Dr. Felix Trombe

This is the only active solar collector we know of that you can walk into—if you can stand the heat. Located in a ranch-style house in a Delaware beach town, it consists of a large skylight opening into a narrow attic room lined with mirrors that concentrate the sun's rays onto a flat-plate collector on the north wall. John Jankus, who lives by himself in the house, sometimes takes a chair to sit in the collector and sunbathe. Since it's made of Plexiglas, the window lets in ultraviolet rays.

Although his field is accounting, John's interest in solar energy dates back to the days when he was a kid and used a magnifying glass to burn ants.

Based on this background of childhood atrocities, John grew up a firm advocate of solar heating.

Although he denies being the chicken-little type who says the sky is falling, he has always felt that energy was eventually going to be a big problem, contrary to the way a lot of people still feel. When the University of Delaware built a solar home in 1973, John toured it the first week after completion. Together with his father, a builder, he decided to design a solar home but wanted something other than the usual flat-plate collector. An uncle read in the *New York Times* that a patent for a "pyramidal optics" system had been issued to

Gerald Falbel and assigned to Wormser Scientific Corporation, a small consulting group in Connecticut.

Reflectors

In a pyramidal optics collector the reflectors consist of plywood covered with aluminized Mylar. They form a pyramid-like optical concentrator—hence the name. The wall and roof mirrors are stationary but the floor reflector is a movable flap that governs the amount of light admitted. A winch controlled by a clock motor raises or lowers the flap to change its position seasonally. Moving about once every two weeks, it changes a total of 5 feet during the year. By December 21 it is all the way down at a 10° inclination to admit as much sunlight as possible; by June 21 it is all the way up to 40° to close the window for summer. These adjustments can be made manually if desired, thereby reducing installation costs.

On the north wall, the absorber consists of a sheet of aluminum fused to a sheet of copper with channels for the water to flow through. The concentrating system reduces the necessary area of flat-plate collector and associated plumbing by a factor of four, thus lowering costs by replacing expensive absorbers with cheaper reflectors. The use of a strategic metal like copper and an energy-intensive material like aluminum is minimized.

A pyramidal optics collector, being an integral part of the roof, is less obtrusive than a standard flat plate, and therefore to some people is more esthetically pleasing. Although the absorber must be angled at the inclination dictated by latitude, the angle of the Plexiglas depends on the roof pitch. The roof can have a low slope if desired. If an attic has rafters rather than roof trusses, the system is suitable for retrofits.

Since the absorber is inside, it is protected from weather. Maintenance is minimal, involving annual cleaning of the floor reflectors. For their first cleanup John lifted them out, carried them out-

---

TECHNICAL DATA

Owner-designer-builder: John Jankus, Delaware

General Features

Latitude: 39° N
Degree-days: 4,900
Insolation: 155
Heated Area: 1,400 ft^2
Year of completion: 1976
Insulation: Walls: 3½" fiberglass and 1" Styrofoam
      Roof: 10" blown mineral wool
      Foundation: 6" fiberglass
      Shutters: ½" louvered wood
Orientation: S
Solar system: Active liquid drain-down (pyramidal optics system)

Collection System

Collector: 370 ft^2 (window), 136 ft^2 (absorber)
  Manufacturer: Site-fabricated by Wormser Scientific Corp., Stamford, CT
  Angle: 40° (roof), 55° (absorber)
  Cover: Single layer ⅛" Plexiglas (window)
  Absorber: Copper RollBond ®, selective coating
  Pump: 1/3 hp (6½ gal. per minute)

Storage System

Container: 8'×5'×6' rubber-lined concrete septic tank
Material: 850 gal. water

Location: Utility room on E end of first floor
Insulation: 3" sprayed urethane

Distribution System

Forced air in 4 modes: 1) When storage is above 85° F., ¼-hp pump triggered by thermostat moves the water to a heat exchanger. Heat pump fan blows heated air through ducts to house. 2) Below 85° F., another ¼-hp pump sends water to heat pump's compressor, operating in its water-to-air mode. It upgrades the solar energy, delivering 95° F. air. 3) When storage drops below 40° F., auxiliary electric resistance takes over. 4) For summer air conditioning heat pump functions in its air-to-air mode with an outside condensor unit for dumping heat.

Auxiliary System

Backup: 40,000-Btu electric resistance
Fuel consumed: $110 at 4½¢/kwh average (electric resistance and heat pump)

Domestic Hot Water

40-gal. tank submerged in storage tank preheats for electric water heater.

Costs

House: $30,000
Solar: $11,500
Operating costs: $15 a year for pumps and controls

John Jankus at door into collector end of attic. At right, interior of the attic. Plexiglas window on left, liquid collector with integral channels on right. Movable floor reflector adjusts to angle of sun.

side, and hosed off the dust. Next time he'll probably leave them in place and use a dustrag.

John keeps his thermostat at a fairly constant level for the sake of monitoring equipment. During his first year in the house, he dropped it to 60° F. on workdays and raised it to 68° at night and on weekends. He then joined the staff at Wormser Scientific Corp. and began commuting between Delaware and Connecticut, spending half the

John Jankus in dining area.

week in each state. Again, when away from home he moves the thermostat to 60°.

Performance

The Jankus house is well insulated and infiltration is reduced by a layer of Styrofoam. Windows are triple glazed. Wormser estimates that without these improvements a house of the same size in the same location would have an electric heat bill of at least $600. The first winter John's heat bill was $110, but the heavy insulation is credited with $300 of the $500 savings. Solar heating of domestic hot water saved about $50, making a total of $250 from the solar system by itself, or $550 from the combination of solar and extra insulation.

Using the $550 figure and an estimated duplication cost of $8,500 (heat pump not included), Wormser projects a "positive cash flow" in the sixth year if fuel costs escalate at 12 percent a year. On this basis the system is then completely paid for by the tenth year, and $25,000 in savings accumulate by the twentieth year. Current predictions about the rate of fuel escalation vary anywhere from no higher than the rate of general inflation up to a frightening 16 percent.

JANKUS HOUSE: SAVINGS FROM SOLAR SYSTEM—20-YEAR LIFE CYCLE

Initial Net Cost: $8,500   Initial Savings: $544/year

Fuel Escalation: 12%/year   Solar System Loan: 15 yr.   9%

Year	Escalation Factor	Saving From Solar	Payments on 15 yr. 9% loan $8,500	Net Cash Flow	Cumulative Savings
1	1.0	$544	$950	−$406	−$406
2	1.12	609	950	−341	−747
3	1.25	682	950	−268	−1015
4	1.40	764	950	−186	−1201
5	1.57	856	950	−94	−1295
6	1.76	959	950	+ 9	−1286
7	1.97	1074	950	+124	−1162
8	2.21	1203	950	+253	−909
9	2.48	1347	950	+397	−512
10	2.77	1508	950	+558	+ 46
11	3.11	1690	950	+740	+786
12	3.48	1892	950	+942	+1728
13	3.90	2119	950	+1169	+2897
14	4.36	2374	950	+1424	+4321
15	4.89	2656	950	+1706	+6027
16	5.47	2978		+2978	+9005
17	6.13	3335		+3335	+12,340
18	6.87	3735		+3735	+16,075
19	7.70	4183		+4183	+20,258
20	8.61	4685		+4685	+24,943

But if Wormser had omitted the $300 savings from insulation, the time intervals would double: positive cash flow in the twelfth year and payback in the twentieth. Thus even against electric heat, a high-technology system is not cost-effective.

Comparison Planned

The pyramidal system is just one of Wormser's sidelines. The company has also installed evacuated-tube air conditioners on banks in Connecticut and New York, retrofitted an office building with flat-plate collectors, and done research on reflective surfaces and control circuitry. A HUD grant is providing Wormser with the opportunity to build three test homes side-by-side, identical in every way except for the type of solar system used. All three will embody some passive collection, but one will have an active liquid collector, another a Sunworks air system, and the third pyramidal optics. Wormser is predicting that all three will perform about equally, providing 71 percent of the space heat, with an additional 17 percent coming from passsive. Surprisingly, they are also predicting that all three will have a similar price tag. Since a pyramidal optics collector is assembled at the building site, its initial advantage in material costs is made up in the field by labor and logistics problems. Specialty items like Plexiglas or Mylar have to converge on the site in a certain time sequence. If any one item falls off schedule, it delays the whole project.

Wormser hopes to get the price down but is not sure whether it can compete with the large corporations. Whether it's photovoltaics, collectors for space heating, heat pumps, or major components, John believes we're going to see domination of the solar industry by manufacturers like General Electric. "The big companies have the money, the resources, and the name." Sears, he says, is negotiating with some large collector manufacturers and will be selling a hot water system "supposedly by next year."

John's collector has received its share of tinkering. Expansion from heat loosened the silicone

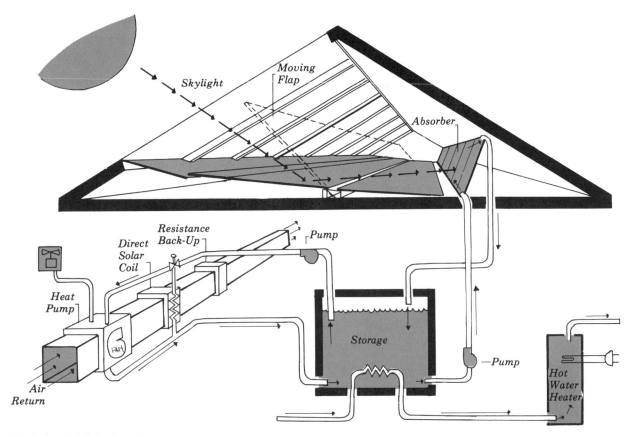

The Pyramidal Optics solar system.

Courtesy Wormser Scientific

caulk used to seal the Plexiglas window and permitted leaks to form. When the house, which sits in an open field, is hit by coastal storms, the Plexiglas flaps noisily. Even when there's no wind, it creaks as it expands—an eerie sort of sound. Worst of all, it's flammable. The Delaware house is in a

Heat pump, heat exchanger, and ductwork in utility room on first floor of house.

rural area where building codes are not particularly restrictive, but in urban areas a Plexiglas skylight of this size would never pass code. Wormser has decided to replace it with low-iron tempered glass.

Building back in 1976, John had problems with banks and property taxes. The bank required him to outline a procedure for converting the house to conventional heating in case the solar collector didn't work. And John is understandably upset about an extra $50 a year tacked on his property taxes. Though most states have now done so, Delaware has yet to enact a law prohibiting tax increases for solar improvements.

Sizing the Heat Storage

A compact solar system is one of Wormser's objectives. If the storage is compact, it becomes feasible to put it on the main floor, which is important wherever basements are impossible. An 850-gallon concrete septic tank is located in the utility

215

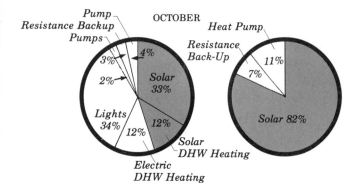

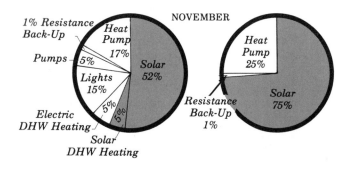

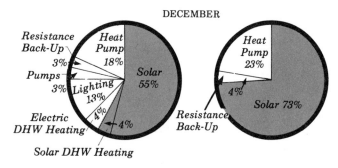

Total building energy (left) and energy sources for space heating (right), October to December, 1976.

room of the Jankus home. The idea is to make solar heating as much like conventional heating systems as possible. Wormser believes that for solar heat to be accepted it has to be both compact and maintenance-free. A repairman once told John

that the average homeowner will not bother changing furnace filters. Often heat pumps and air conditioners break down simply because there's so much dust in the filters that air can't get through. "If they can't change their air filters," John asks, "what are they going to do?" He may be right that Americans have become this lazy, but the fact that New Englanders in droves are turning to wood stoves to fight high oil prices would indicate that he's selling the people short. Perhaps when the wallet is involved we aren't opposed to a little inconvenience.

Furthermore, compactness in terms of small storage means reducing carry-over of heat to such an extent that there seems little point in having an active system. John reports that from fall until early winter, his tank temperature stays between 110° and 130° F., providing 100 percent solar heating. The same is true from March until summer. But between December and March it drops off quite abruptly because the heating load has increased but the storage is exceptionally small. A day of cloud cover and the heat pump drains the tank heat down to 40°. Why not just go with passive in that case? It too has a carry-over of one day but is a lot cheaper.

John believes that Americans will not accept the temperature fluctuations involved with passive heating. But with our pocketbooks under seige, perhaps we can get used to a few ups and downs. After all, we read that before central heating was invented, our great-grandparents chipped ice out of the kitchen water bucket on cold mornings. Although passive solar heating doesn't mean that kind of hardship, have you ever considered that it might be unhealthy and perhaps a bit unnatural to maintain monotone temperatures in our living spaces no matter what's happening outdoors? Do we need to live in incubators?

Solar Satisfaction

The sunrise never failed us yet.
Celia Laighton Thaxter

"We are totally satisfied with our solar heating system," say Dorothy and Bob Rusher. They expected it to work, of course, but were not prepared for the kind of comfort they have encountered. It's largely due to the lack of drafts, which means there is no need to turn the thermostat up from their accustomed 68°. Even Bob's mother finds it unnecessary to put on a sweater in the evening.

But many people interested in solar energy are young and don't have the money to do what the Rushers have done. Their home is the culmination of a lifetime of saving and experience. It's the fifth house they have built, and, they hope, the last. Still, Bob fears that they're doing too little too late: "We Americans are not used to conserving and it comes hard to us. The rest of the world is far ahead; in Japan and India, for example, you see solar water heaters on rooftops everywhere." Perhaps it

is because sunshine is so commonplace in our lives that we ignore it as a practical source of energy.

With the help of designer James "Jay" Lynch, the Rushers spent two years planning their new home outside Montrose, Colorado. It was a long process in part because Lynch lived 100 miles away, but also because of the caution with which they approached the project. Formerly a chemist with duPont in Delaware, Bob went about the design with accustomed thoroughness. The Rushers, whitewater canoe enthusiasts, had moved West in 1973 to enjoy the mountains and get away from crowds. Dorothy worked as a secretary while Bob took up real estate sales. Quite by accident he found a kindred soul in Lynch, who came to the real estate office searching for farmland.

Together, the Rushers and Lynch examined the possibilities offered by their site. Dorothy wanted

217

a view of the spectacular San Juan Range directly to the south. At first Jay rejected the idea of the classic picture window. Then when analysis indicated that the Rushers could meet half their heating needs through passive means, he included two in the design. Additional windows on the east and west admit heat during the morning and afternoon hours. With the exception of the picture windows, triple glazing is used throughout. Direct gain is sufficient to heat the entire structure during daylight hours.

Passive Heat Underrated

Too little thought is given to designing houses to get all the passive gain possible; the Rushers found that people don't even understand the concept. Dorothy has noticed that in her neighborhood housewives close the drapes the minute the sun starts to come in the windows, even in winter. Yet passive gain is permitting the Rushers to derive 75 percent of their heating from the sun.

The roof overhang is designed to reduce glare and prevent summer sun from reaching the interior of the house. Old Sol begins to creep in the front glass in the middle of September, and by the equi-

Bob Rusher in living room, with kitchen behind.

nox, when it is really needed, reaches to the back walls. "I'm sure there'll be some fading, but I'm not going to worry about it," Dorothy says. Grouped against the north wall facing the view, the furniture is usually out of the limelight.

TECHNICAL DATA

Owners: Dorothy and Robert Rusher, Colorado
Designer: James Lynch
Solar consultant: Joe Costello

General Features
Latitue: 38° 30' N
Degree-days: 6,800
Insolation: 200
Heated area: 2,300 ft^2
Year of completion: 1977
Insulation: Walls: 6" fiberglass and 5-mil polyethylene film
      Roof: 14" fiberglass and 5-mil polyethylene film
      Foundation: 2" high density Styrofoam
Orientation: S
Solar system: Hybrid—passive direct-gain and active air

Collection System
Passive collector: 150 ft^2 double and triple-glazed windows
Active collector: 260 ft^2 airHair™ (14 panels)
   Manufacturer: Tritec Solar Industries, Durango, CO
   Angle: 60°
   Cover: Double layer of tempered glass
   Absorber: 4" glass fiber matrix, nonselective coating

Storage System
Container: 8' × 5' concrete bin
Material: 10 tons of 1½"-3" dia. rocks
Location: Under entranceway between garage and house
Insulation: 4" high density Styrofoam

Distribution System
½-hp blower circulates air through rock bin at 700 cfm, through duct system to registers.

Auxiliary System
Backup: 12-kw electric in-duct heater and fireplace
Fuel consumed: $85 at 3½¢/kwh average, 1977-78

Domestic Hot Water
Coil in rock bin preheats for 30-gal. electric water heater.

Costs
House: $65,000
Solar: $8,000
Operating costs: $50 per year

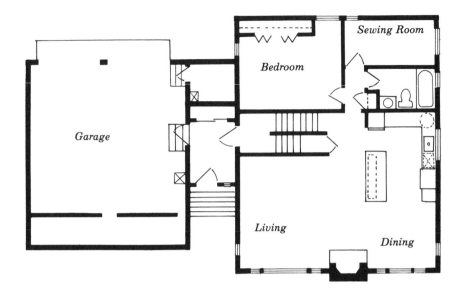

UPPER FLOOR PLAN

LOWER FLOOR PLAN

It is the east and west living room windows that present the problem, because the eaves are narrow. Rubber-lined drapes don't block the heat of the July sun. Last summer there was no overheating but this year temperatures have been higher than normal, and Dorothy has been uncomfortable. They first installed exterior shades, manually operated from inside. That didn't provide the desired comfort so they finally installed an air conditioner.

Another source of unwanted heat is the black roof. At the time they built they had no particular reason for selecting black over white. Actually, as Bob points out, it should be black in the winter and white in the summer, but that's hard to accomplish. He would like to see an aluminized Mylar film that could be rolled down over the roof. Lacking such an invention, the black surface acts like a magnificent collector, so much so that the Rushers are considering reroofing in a lighter color. They have already added more soffit vents to draw heat away from the attic. Together with a larger ventilating fan, these may enable them to avoid the roofing job.

Energy-Efficient Construction

The house should now function as well in summer as it does in winter since it is well insulated. Special roof trusses permitted the extension of R-37 insulation in the ceiling to the exterior of the walls. Below-grade concrete walls and the slab were insulated with Styrofoam M, as were the window and door headers. "We have $1000 worth of insulation buried in the ground," Bob says, "but it's worth it." The outer corners of the house were framed so that they could be insulated, and even the wiring was snaked along the sill plates through holes drilled in the bottom of studs so as to avoid compressing the fiberglass. During the construction Bob and Dorothy spent evenings stuffing insulation in the little nooks and crannies that ordinarily would not have been filled, such as around the electric boxes. The insulating features have reduced the heat requirement of the house to a fairly low 8,500 Btu per degree-day.

The handsome brick fireplace, while it is used mainly for enjoyment, is also designed for energy-efficiency. Located in the living-dining area between the picture windows, it has glass doors and a double-walled firebox, around which room air is circulated and warmed in passage. Outside air brought into the combustion chamber reduces drafts across the floor. The house is so tight that if the damper in the outside air duct is closed, the fire won't burn. Only one problem has turned up in the operation of the fireplace: exterior air rushing in causes too much turbulence in the fire. Bob believes that if the intake pipe were 6 inches in diameter instead of 4, this difficulty would not exist.

Choosing Collectors

While the winter sun is heating the house passively during the day, fourteen collector panels mounted on the garage are gathering solar heat and putting it into storage for use at night. Several commercial collectors were investigated before the choice was narrowed to the Solaron and Tritec collectors. "Then the Solaron prices went up, making Tritec $2,000 less expensive, and that decided it for us," Dorothy says.

The collectors' airHair™ absorber, made by Owens-Corning, is a glass fiber matrix much like a furnace filter in appearance. A blower washes the

Section A-A'.

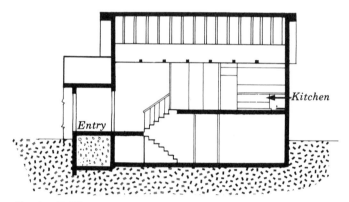

Section B-B'.

heat off the airHair and carries it through glass fiberboard ducts to the rock box underneath an air lock vestibule. The storage chamber, filled with washed river pebbles, can retain enough heat for a night and one or two cloudy days. The short carry-over is due to the small size of the chamber.

Plenty of Heat

A glance at a digital readout in the living room told us how the system is doing. On the day in July when we visited, the collectors read 227° F. at 4 P.M. while the storage chamber was 130°. The system is supposed to turn off during the summer when the storage temperature reaches 90° but for some reason was not doing so that day. "When it does turn off, won't it deteriorate from the high temperatures of stagnation conditions?" we wondered. "There's nothing to deteriorate," Bob replied. According to the manufacturer the collector is also resistant to thermal shock: accidental turn-

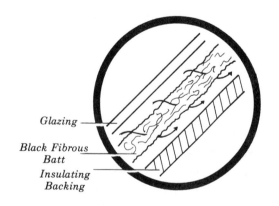

The Owens-Corning airHair™ absorber.

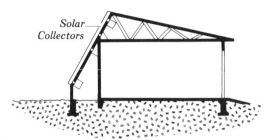

Garage cross-section.

ons will not break the glass or harm the fiberglass matrix.

The Rushers have absolutely no complaints about their solar system. The installation was conducted by Tritec under the supervision of president Joe Costello, a former physics professor who dropped out of academia. On the few occasions when there have been problems, mostly electrical, the company has sent someone out within 24 hours. Once there was an air leak in a duct where the tape came off. Dorothy emphasizes that little things like that are nothing compared to the troubles they've had with one of the showers.

Bob reports that on a good collecting day in winter the air coming down the collectors will read

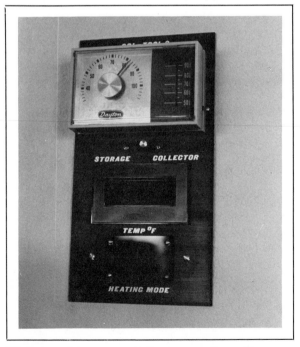

Digital readout—center indicates storage or collector temperature (227°) as well as heating mode at bottom (° for off). Living room temperature on a July afternoon was 82°.

as hot as 270° F., though 250 is more usual. "You might think that when a cloud comes over, the collector would shut down, but that's not so," he says. "It's amazing how much heat it can accumulate even when the sky is heavily overcast." Last winter was very cloudy but electricity for the back-up cost only $14 to $16 a month.

True Energy Cost

The Rushers claim that the cost of the active solar system will be amortized in five to seven years, depending on the rate at which utilities raise their prices. According to Tritec, however, the best single gauge of solar value is the True Energy Cost (TEC). The formula for determining the TEC is:

$$\frac{\text{Total Cost of System}}{\text{Yearly Energy Received} \times \text{Lifetime of System}}$$

$$\text{Plus} \quad \frac{\text{Yearly Operation Cost}}{\text{Yearly Energy Received}}$$

The Rusher house yields the following values:

$$\text{TEC} = \frac{\$8,000}{43.35 \text{ mil. Btu} \times 30 \text{ yrs.}} + \frac{\$50}{43.35} = \frac{\$7.30}{\text{Btu used}} \text{ per million}$$

This calculation includes the energy derived from the passive gain. Yearly operating costs are based on fan motor replacement every ten years and one complete replacement of all mechanical control elements during the useful lifetime of the system. The cheapest alternative in the area, natural gas, costs $5.10 for each million Btu used, assuming the furnace is efficient and well-tuned. Electric resistance heating costs three times that much. If we exclude the 21.68 million Btu of passively derived heat, we find that Btu gathered by the active system cost $13.45 per million. While that is cheaper than electricity, the passive is clearly a better buy.

Bob feels that it is still too early to be able to justify active solar heating on cost savings alone. Equally important considerations are reducing air pollution and conserving fossil fuels for use as chemicals. But it's a rare person who will put in

View of house from northeast.

solar collectors only because he's thinking of what needs to be done. The federal tax incentive should help, but Bob was not the type who sat around waiting for the government: "What I tell people is that they can't sit on their duffs complaining about Uncle Sam not solving this problem and at the same time attack big government."

Quit Stalling

Bob suspects that the reason solar heating hasn't taken off as fast as it should is that there's no way for the big corporations to make a killing on it. The most effective solar-heating systems are tailor-made to suit each house, and the installation is labor-intensive. It takes dedicated and knowledgeable people to develop solar heating commercially. Bob would like to take his message of solar satisfaction into households across the country: "We Americans have the most effective means of communication available today in every family's living room. If the 'boob tube' were turned to good purposes, we could educate a new generation of energy-conservers in no time." All we have to do is quit stalling—searching for exotic scientific breakthroughs—and use what's coming in the window.

The Sun Machine

A house is a machine for living in.
Le Corbusier

"I think a piece of machinery sitting in a field can be beautiful because of the contrast between high technology and the primitive. To me the most important thing is how well that machine works. If it functions well, it will be beautiful and harmonious with the environment even though it may look like a spaceship."

Thus Mike Jantzen defends his home when visitors say it looks unnatural in its woodland setting. It may resemble something from another galaxy, but Mike believes that some of today's most esthetically pleasing designs are emerging from the

South view of the Jantzen home, showing two Plexiglas windows. A domestic hot water collector is directly beneath the central set of windows.

space program. A rocket, like a tree, is designed to do what it does as directly as possible—to get a job done well. Mike thinks that the American people are due for a change in their conception of beauty, because too often we praise buildings whose beauty is a skin-deep facade hiding a rotten core. His house, unlike the typical American dwelling, does not pollute the environment. Its beauty is more than surface-deep. It is, Mike says, an energy-efficient machine.

Both Mike and his wife Ellen are artists, Ellen a commercial artist with side interests in weaving, gardening, and writing, and Mike a sculptor who refers to himself as a conceptual artist. Conceptual art, a movement of the '60's, sees the information conveyed by a work of art as more important than the visual elements so that any material could be chosen to put across a given idea. For Mike, experimenting with materials evolved into experimenting with structures, including a solar vacation house built in 1975 and the "solar spaceship" in 1977.

An hour's drive east of St. Louis, Mike and Ellen's house is located in a lightly populated section of Illinois farmland. Their land, consisting of seventeen acres of oak and hickory woods dotted with apple orchards, was Mike's birthplace. He and Ellen bought it from Mike's father, then in twelve months did all the construction work on the "machine" themselves. Most of the building materials were donated in return for publicity rights, so that the Jantzens' actual investment is about $7,000. Apart from some supplementary income from designing furniture, teaching, and doing carpentry, they are nearly self-sufficient, producing most of their own food and energy.

The Jantzens prefer to direct attention, not to the solar aspect of their home, but to its total energy efficiency. Electricity and water use, as well as environmental impact, are carefully controlled. Energy is used and reused, and even human energy is saved. Down to the tiniest details the word is convenience: a desk on wheels with a built-in seat; hanging pockets that resemble shoe caddies for storing small items; numbered recycling containers for separating glass, aluminum, steel, and plastic.

The Jantzens would like to see less preoccupation with solar hardware and more with creating new life-styles. Mike is all for technology, but within limits: "If everything is done for you, I think you become deadened, desensitized to things around you." The Jantzens are trying to take advantage of technology while at the same time living harmoniously with nature.

Walking through the Jantzen house is like taking a course in energy conservation. In the living area electricity use is minimized by flourescent

---

TECHNICAL DATA

Owner-designer-builders: Ellen and Michael Jantzen, Illinois

General Features
Latitude: 38° 30' N
Degree-days: 4,500
Insolation: 150
Heated area: 2,000 ft^2
Year of completion: 1977
Insulation: Walls: 3" urethane sprayed with fire retardant
Roof: 4½" urethane with fire retardant
Foundation: 2" urethane and 1" sprayed foam
Shutters: 1"-3" urethane
Orientation: S
Solar system: Hybrid—passive direct-gain and indirect gain and active air

Collection System
Passive collector: 184 ft^2 (living room windows) and thermosiphoning air collector
Active collector: 320 ft^2 (not yet operating)
Angle: 53° to almost 90°
Cover: Single layer Kalwall Sun-Lite®
Absorber: Corrugated steel siding, 1" fiberglass board and 3" sprayed foam

Storage System
Container: 12' × 12' brick floor in living room and two 11½' × 8' × 1½' concrete block bins

Material: 15 tons 4"-6" diameter rocks
Location: Crawl space beneath living room
Insulation: 4" foam (sides), 2" (bottom)

Distribution System
Natural convection through vents at bottom and top of air collector. Two blowers will be mounted in ducts to pull air down into rock storage. Floor registers will permit heat to rise into house.

Auxiliary System
Backup: 75,000-Btu wood stove
Fuel consumed: 2½ cords

Domestic Hot Water
Collector: 64 ft^2
Angle: 40°
Cover: Single layer Kalwall Sun-Lite
Absorber: Corrugated steel siding with attached ½" copper tubing
Storage: Synthetic oil transfer fluid pumped from collector through tubes around 40-gal. tank. Boosted by electric in-line water heaters.

Costs
House: $72,000 (estimated value)
Active solar: $500

North side of the house. Corrugated steel siding on curved surfaces. Material on flat surfaces is stone aggregate epoxied to plywood. The doors are an air lock entrance, a pantry, and a storage compartment for dry food. The section for dry food such as grains, rice, beans, and lentils is insulated from the house and vented to keep it cooler.

fixtures, which are used in reading lamps. Low-wattage incandescent bulbs are situated wherever lights are turned on and off frequently, as in the pantry. Where possible, the Jantzens installed pull chains instead of switches to limit the amount of wiring while light-colored walls reduce the need for artificial lighting.

Reliance on electricity from fossil fuels is further lessened by means of an "energy cycle," a device that lets the cyclist tone up while charging batteries or grinding grain. As they pedal, the Jantzens read or watch television, at the same time producing power for lamps, TV, and a sound system which consists of a four-speaker D.C. car stereo that runs for months on a 12-volt battery.

In the kitchen and bath, high-pressure spray nozzles reduce water flow, permitting a gravity-feed water system. To minimize use of the electric pump, once a week water is pumped from the well to a 120-gallon storage tank on the second floor. In the bathroom hot water is conserved by an insulated shower stall. A Mylar curtain ensures that the stall warms up quickly, and a towel rack inside permits drying off before reentering the bathroom, making a ceiling heater unnecessary. A

Ellen: "In an energy-efficient house you feel a part of the whole system, rather than apart from it."

225

hand-held hose replaces the conventional shower head, because rinsing off takes less time if the water doesn't have to run down your body. A shower with this system takes about 2½ gallons.

Domestic Hot Water

The solar water heater is angled for summer instead of winter sun. Mike plans to hook up a heat exchanger to the wood stove for winter preheating in order to reduce use of the electric in-line water heaters. An in-line heater operates only when water comes through the line, instead of using energy to maintain a 120° temperature in 60 gallons stored in a tank. Hot water can never be exhausted as it can with a conventional heater, and as long as the flow is kept low, the water will get as hot as you could possibly want.

The 16-foot-long solar hot water collector is made of corrugated metal siding, installed horizontally. To minimize soldering Mike made a nest of steel wool in the valley of each corrugation and stuffed copper pipe down into the wool. Clamps help hold the tubing in place. At the end of each corrugation the pipe exits, makes a narrow turn, and reenters. Synthetic oil is pumped through the pipe and to coils surrounding a 40-gallon holding tank.

Treatment of Waste Water

Gray water from the shower and kitchen sink is piped to an underground container for eventual reuse for watering flower and herb beds. A compost bin is located above the gray water tank to allow heat rising from it to speed composting action. If you're interested in this system, check your local sanitation code as a small leach field is usually required for gray water.

Because of odor, the Jantzens had to replace an anaerobic or liquid-system composting toilet with an aerobic system. The new Mullbank uses a continuously running blower for ventilation and a heating element to keep temperatures at an optimum level for composting. It uses the equivalent of three 60-watt bulbs, but the odor problem is gone. Although small flying gnats remain a nuisance,

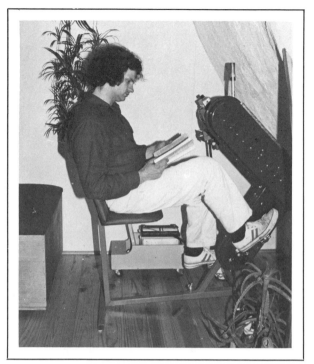

Mike on the "energy cycle."

Herb bed air freshener and natural insect and rodent repellent.

One of the studios. Unfinished at present, the spaces above the studios will become a guest sleeping space and a darkroom.

an organic spray can be used to control insect populations.

With a microwave oven, a refrigerator that uses no more power than a 75-watt bulb, and the in-line water heaters, the Jantzens use only 250 kilowatt hours per month—well below the minimum billed by their power company. A freezer, currently out of order, adds another 100—less in the winter because it's kept in an insulated pantry. The freezer's waste heat is vented into an air lock entrance while the refrigerator's goes through a screen into a cabinet above, where bread is put to rise. The pilot light in the gas stove provides heat for drying fruits and vegetables.

The sheer quantity of ingenious energy-saving ideas is what's appealing about the Jantzen house, not to mention novel applications of old ideas. For instance, they've revived the old-fashioned bed-curtain to create a warm mini-environment that permits the rest of the house to be kept cooler at night. Made out of space blanket material, the curtain forms a tent that efficiently retains body heat. A similar curtain down the middle allows one person to read while the other sleeps.

Built on a flat ridge top, the house is surrounded by woods carpeted with May apple and wild mush-rooms. Along with the air lock and pantry, compartments for dry food, firewood, and tool storage form an insulating buffer on the north side of the house. The doors, like many closets in the house, are numbered for convenience: "If Ellen wants something, she just tells me which door," Mike says.

Windows and Ventilation

The building's windows are stationary to reduce wind infiltration; instead, ventilation is accomplished through vents. Plexiglas bubbles on the south side are fitted with movable shades, like eyelids, that collapse against the wall in winter. On the north two porthole windows, small to prevent heat loss, are fitted with wide-angle lenses to increase the viewing area. Plastic fresnel lenses, available at novelty stores, distort slightly but create an interesting effect.

All the windows, including the portholes, are equipped with interior shutters. As you go up in the house their thickness increases for better heat retention. Consisting of foam sandwiched between white Masonite panels, the larger, heavier shut-

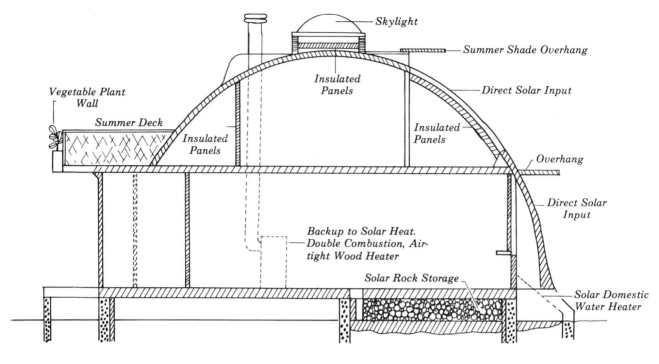

CROSS-SECTION

ters slide closed on tracks. Mike now realizes that he should have bought steel rather than soft aluminum tracks. The aluminum is becoming torn by the rollers, causing the shutters to catch.

Rotating ventilators on the roof are opened in summer to pull air through vents located near the lowest points of the living spaces. Herb beds near the air intake vents at the corners provide a natural insect and rodent repellent. Aromatic herbs like mint and tansy were selected because they repel flies and mice while freshening the air. Other vents open into the crawl space with companion registers inside the house, creating a cool airflow in summer. "When we have a north breeze," says Mike, "it's almost like standing in front of an air conditioner." Eventually the crawl space warms up, but always remains cooler than the outside. Shaded by a dense canopy of deciduous leaves, the house stays fairly cool, yet in a climate as hot and humid as that of the Mississippi River Basin, Mike believes that it may make sense to build underground.

With plastic windows and steel siding, the house makes heavy use of energy-intensive materials. Since building it Mike has been reevaluating materials but still argues that it's better to use plastic, a petroleum-based product, in a house than to burn up our oil in automobiles. The corrugated steel is a prepainted siding that is relatively inexpensive, maintenance-free, and structurally strong so that the number of supports can be reduced. The supports are preformed arches, which dictated the shape of the building and made construction quick and easy.

On each end of the house is a studio, which is a completely independent structure bolted to the main building. If the Jantzens ever wish to expand the central core, a tractor can simply pull the studios out onto new foundations. Because the connecting doors are insulated, these spaces can be closed off and allowed to freeze. The Jantzens in fact heat only part of the building's 2,376 square feet. The principle of zone heating is carried even further: the ceiling of the first floor is insulated with 6 inches of fiberglass in order that the upstairs bedroom can remain unheated when it's not in use.

Air Collectors

Two air collectors, currently operated by thermosiphoning, flank the central south-facing windows. Here the corrugated siding was installed

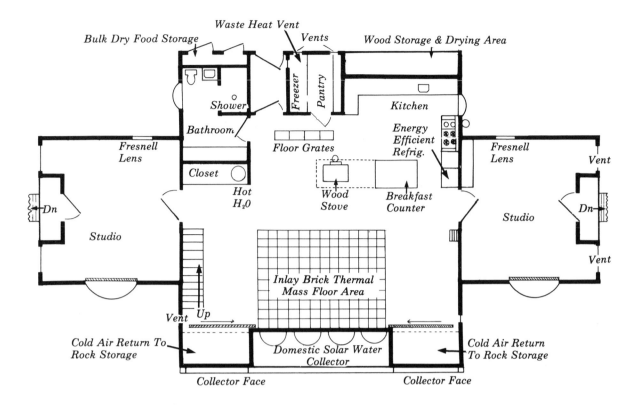

FIRST FLOOR PLAN

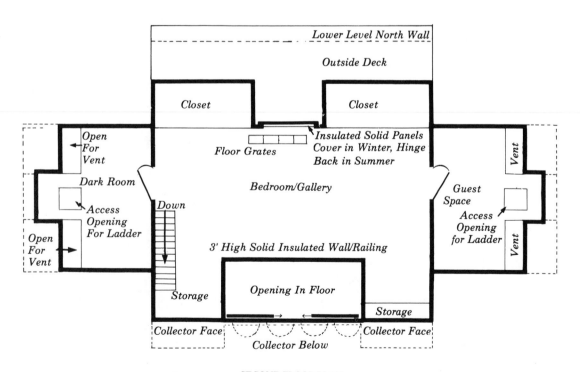

SECOND FLOOR PLAN

229

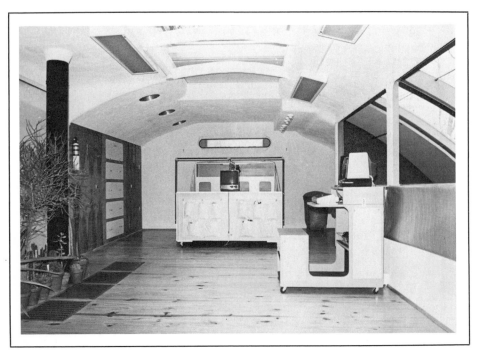

Bedroom with balcony overlooking the living room. Shoe caddy pockets are attached to the foot of the bed for storing small items. The bedcurtain pulls over the pipe frame.

vertically. Following the S-curve of the wall, part of the collectors are not at an optimum angle, a lapse in efficiency out of character for the Jantzens. Once the system is complete, separate collectors will mean extra hardware—two blowers—but will provide flexibility. When more sunlight is striking one collector, the other can be shut down. Mike plans to replace the Sun-Lite cover, which deforms so badly when hot that it touches the absorber. While getting the cover to fit tightly on the curved surface was difficult, Mike points out, and we agree, that he has yet to see a collector with Sun-Lite that doesn't wrinkle to some extent.

The system is still passive since the blowers and ductwork are not yet installed. Although winter temperatures sometimes drop well below zero, the house is always 70° F. when the sun is shining. Mike and Ellen once ran a test: after thirty-three hours with no wood heat the house was 52°. Normally they keep it between 65° and 68°, and 60° at night. Even in the coldest weather the wood stove is filled only once daily. A cart that slips beneath a kitchen counter holds a three-day supply of wood. By the end of March the Jantzens are only burning trash in the morning to take off the chill.

Neighbors up the road with a house of the same

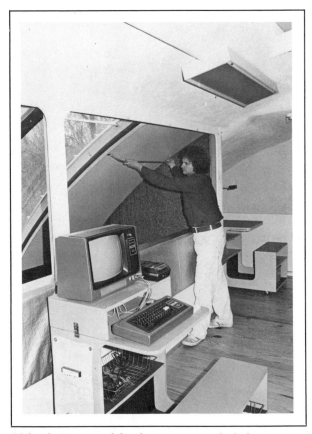

Mike closing one of the shutters next to the balcony overlooking the living room. Computer sits on a desk on wheels with a built-in chair.

size have electric bills of $600 per season despite relatively inexpensive rural electric service. Heating with oil would probably cost the Jantzens more than $300, so that even if they bought their firewood, they would still save $200 a year. Their solar investment consisted of $160 for the Sun-Lite, $90 for rock storage, and about $250 for hardware and insulation. The absorber added nothing since it also serves as the skin of the building, and concrete block for the storage bin does double-duty as foundation walls and floor supports.

Because the crawl space is only 2 feet high, the rock storage had to be shallow and long, which will mean a lengthy run for the solar-heated air. Mike tried to make up for consequent resistance by using larger rocks. The building has worked so well without the active collector that Mike says if he were to design again, it would be a totally passive system.

Computer and Compost

Tailored for the Jantzens' life-style, the house is epitomized by the coexistence of the compost pile and a computer that was donated for monitoring performance. Carrying recycling to its technological climax, Mike is planning to reuse the computer's electric signals to make music, create games, and produce visual patterns on a television screen.

Although we may not agree that a house should look like a solar machine, the Jantzens have a point: we should be flexible enough to make some changes in our esthetic preferences for the sake of energy conservation. As for making technology compatible with a natural way of life, the Jantzens seem to have gone a long way toward demonstrating that the computer and the compost pile can get along.

Afterword

After visiting nearly a hundred solar homeowners in the course of our research, what do we conclude? The inescapable finding is that solar energy is not only here now, it is economically viable in almost every situation.

With the exception of one or two unfinished houses, such as that of Noel "Paul" Stookey, all the systems described in this book have been proven to work, both in terms of providing heat and saving money. But you have probably noticed that the cheaper ones, especially the passively heated homes, tend to save as much or nearly as much in fuel costs as the active systems, at a fraction of the initial investment.

People who claim that more technological breakthroughs are needed before solar heating can become a big-time energy source haven't looked around. Solar systems already exist to fit every pocketbook and esthetic taste. They range from do-it-yourself passive greenhouse to high-efficiency evacuated tube collectors.

How to Choose a System

How, then, to choose from among all the possibilities? In some cases the choice is dictated by the situation: retrofitting an urban dwelling which faces the wrong way, as in Dorothy Maxwell's case, may require an unusual custom-designed system. Ordinarily there are more options. Our recommendation is that you take advantage of passive gain before going to a complex system. Tom Zaugg said it all: "Anybody who builds a building that does not utilize passive solar heating is insulting the earth we live on."

If either of us were to build again, we would definitely choose either a wholly passive system or passive with a fan. The exact type is still an open question. Research comparing direct- and indirect-gain systems has been initiated in different parts of the country, but we know of no definitive results at this time. At least for now the choice is a matter of individual preference. For one of us the experience of living in a home that receives about half its solar heat via direct gain would incline us toward an indirect system—not because of temperature swings or fabric fading, but because of glare. For many people this would not be a significant irritation, but for those whose eyes are bothered, a Trombe wall or greenhouse containing a thermal storage wall would be the obvious choice. Rules of thumb for passive solar buildings have been published in the *Bulletin of the New Mexico Solar Energy Association* and reprinted by Sandia Laboratories.

Retrofits

The real issue lies with retrofitting. Nine out of ten buildings that will be standing in the year 2000 are standing now. Many of these are thermal disasters. The first step always, as Rip Van Winkle points out, is to lower the heat load by lowering the thermostat and by adding insulation, double and triple glazing, weather stripping, and air-lock vestibules. Then investigate the possibilities for adding passive solar heating: open up your south wall with windows, put up a Solar Room with 55-gallon drums for storage, build a do-it-yourself greenhouse with a thermal wall and move heat into your house by natural convection. Or add an air collector with a thermosiphon system.

For some, none of these alternatives will serve. Perhaps your house is too close to your property line to add a greenhouse. Maybe a neighbor's trees block your access to sunlight. There could be an esthetic problem. If all else fails, in other words, you may have to look to your roof for sunlight. If so, once again there are passive choices, such as Norm Saunders's Solar Staircase™ where sunlight through a translucent roof warms containers of water that radiate heat down into the rooms. But if you're reluctant to put water bottles in your attic for fear of sudden showers, you're backing us into a corner where we have to say, Go ahead and do an active system. For large apartment buildings and commercial installations, active collectors are often the best choice, although even there passive techniques can sometimes be employed.

Liquid or Air Collectors?

Suppose an active system is your only reasonable solution. Do you go with liquid or air? Air systems and trickle water collectors are the least complicated and cheapest. In talking with solar homeowners we heard more gripes about the high-tech systems, in part because whenever you invest a lot of money in a solar system, you're more particular about its performance. We have summarized the good news and the bad news for both air and liquid collectors.

Advantages	*Disadvantages*
AIR	
Lower initial cost	Ducts bulkier, more prone to
Less complex system	air leaks and harder to
Less subject to deterioration	insulate
Reversible for air condition-	Blower uses more electricity
ing (using cool night air)	Require more storage space
Adaptable to domestic hot	
water (through various	
special arrangements)	
LIQUID	
Easily adaptable to domestic	Higher initial cost
hot water	More complex system (more
Easier to install pipes	maintenance)
Pump uses less electicity	Subject to leakage and
Smaller storage required	corrosion
Can attain higher tempera-	Heat exchanger necessary
tures	(except with radiant and
More applicable to large	baseboard distribution)
buildings	Danger of freezing

Before becoming immersed in the air-liquid debate, ask yourself if you have done everything possible to make your house energy-efficient. Solar heating must be considered as one aspect of an overall energy-conservative building design, beginning with proper siting, orientation, and insulation. By using the natural protection against wind provided by trees and landforms and capturing winter sun, a house can be relieved of most of its energy burden. Retrofitters can plant trees and place berms on the north, at the same time trimming trees and removing awnings on the south to admit more sun.

An environmentally sound plan begins with a rational square footage. One of the most common mistakes made by homeowners, solar or otherwise, is building houses that are too large. Compactness of design combined with low ceilings makes energy sense. With careful planning of traffic flow and storage areas, most people could be quite comfortable with two-thirds or half the floor space they have. In the past twenty years, although the size of the American family has declined, houses have gotten bigger. In 1950 the average number of square feet was 900, as opposed to 1,500 today. For those who are stuck with oversized relics, zone heating can help. Hot air trapped beneath high ceilings can be redistributed by means of air recyclers.

Air conditioning can generally be avoided through the use of overhangs, balconies, and well-designed ventilation systems. Thorough insulation can reduce both heat loss in winter and overheating in summer. A look at traditional building styles can often provide clues on how to design to fit local conditions in terms of site, climate, and community.

Snares to Avoid

Suppose you've decided on your system: what should you watch out for? Finding the right person to design and build a passive system or install an active collector is the first problem for those who prefer not to undertake the project themselves. Your local contractor may not be familiar with or interested in solar heating. If not, keep looking until you find someone you have confidence in. Read enough about solar heating to insure that you understand your system so that you can check the installation—it belongs to no one but you. (A selected bibliography of particularly useful books that are frequently mentioned by solar owners is included at the end of this chapter.) Collectors may be warranted for materials but if installed incorrectly fail to perform adequately. And doing passive solar heating, as Sara Balcomb said, takes know-how if you expect to achieve peak functioning.

For the owner-builder the most difficult question may be what materials to use. New products are coming out every day but often have unforeseen drawbacks. We ran across several people, for example, who used vinyl stripping on their collectors, only to find that it shrinks drastically. So does butyl rubber caulk, even though the container may say otherwise. Many people are unaware of the difference between polystyrenes in general and Styrofoam, which is a specific brand (see the Fried house). We lack the space to summarize the recommendations passed on to us, but the *Solar Age Catalog* is a good source of additional information.

The problem of sizing solar equipment is a continuing source of debate, especially relating to storage. Some insist that the larger the better, no matter whether your system is active or passive. It is essential to determine precisely what you want your storage to do: if the storage medium is also the heat transfer medium, you cannot tolerate too low a temperature. Keeping the temperature high entails using a smaller storage. On the other hand, if storage is to adequately temper temperature swings, it must be large. Either way, it's most important to insulate the chamber extremely well. In the end cost may place limitations on the size, but do as much as you can afford. Space may be another limiting factor, but if room for formal storage is restricted, look around for alternative ways of incorporating mass. Quite a bit of heat can be stored in water drums or a couple of extra layers of Sheetrock on the ceiling. You should not be deterred from solar heating because you can't have the optimal setup.

Before you start to design a house or do a retrofit, it is important to check your local and state building codes and to investigate property deed restrictions and subdivision covenants. Zoning regulations can also affect your plans; what if in the future someone builds a high rise blocking your sun? Although solar homes are now being funded with loans from the Federal Housing Administration (FHA), the federal restrictions may restrain your creative efforts.

Some people, it seems, were waiting for Washington to legislate tax rebates. In the case of passive systems, which are inexpensive anyway, the significance of rebates is minimal. The Department of Energy actually is doing more to promote solar energy than is often realized. But in comparison to the millions spent on nuclear research, it's merely a mote in a sunbeam. HUD research grants generally go to the big corporations or architect-engineering firms, because only they can afford to do the paper work. In fact we found at a recent conference that even architects are often confused and discouraged by the application forms.

As solar heating becomes more widespread, laws will be required to protect the consumer from fly-by-night retrofitters. But a premature imposition of overly strict performance standards for solar equipment would limit experimentation and raise costs without assuring greater efficiency. Knowledge is your best protection.

Visit Solar Homes

Our final word of advice for those considering solar heating is to visit as many solar homes as you can. Your regional solar energy society or state agency can provide you with the names of people willing to share their experience. Or call the National Solar Heating and Cooling Information Center (1-800-523-2929) for all kinds of helpful pointers. Be aware, though, that solar homeowners may be busy, or reluctant to discuss their problems. At least now you know some of the questions to ask.

Inquire about lifestyle changes. What's it like to live at 65° instead of 70°? (Or better yet, try it yourself.) We in the twentieth century have enjoyed a brief period of fossil fuel abundance. Now we are having to rediscover conservation practices that were common only fifty years ago: planting windbreaks, closing off rooms in winter, donning long johns. Our mothers knew how to achieve natural air conditioning. They closed the drapes to darken the house on summer mornings.

You'll discover that people enjoy manipulating shutters, opening and closing vents, even chopping their own wood as long as solar heat means only 2 cords to chop. As the Johnsons found, gathering around the stove or fireplace puts something important back in family life. Solar heating reawakens awareness of ourselves and concern for our ecosystem. We have just one world. That world is a good world, and it's a good feeling that comes from being at home in the sun.

Selected Bibliography and Suggested Reading

Adobe News, Inc. P.O. Box 702, Los Lunas, N.M. 87031.

Anderson, Bruce. *The Solar Home Book.* Harrisville, N.H.: Cheshire Books, 1976.

Behrman, Daniel. *Solar Energy: The Awakening Science.* Boston: Little, Brown, & Co., 1976.

Bulletin of the New Mexico Solar Energy Association. Box 2004, Santa Fe, N.M. 87501.

Daniels, Farrington. *Direct Use of the Sun's Energy.* New York: Ballantine Books, 1964.

Energy Primer. Portola Institute. Fremont, Cal.: Fricke-Parks Press, Inc., 1974.

Kern, Ken. *The Owner-Built Home.* New York: Charles Scribner's Sons, 1975.

Leckie, Jim, et al. *Other Homes and Garbage.* San Francisco: Sierra Club Books, 1975.

Passive Solar Buildings: A Compilation of Data and Results. Sandia Laboratories, Albuquerque, N.M. 87115. Available from: National Technical Information Service, U.S. Department of Commerce, 5285 Port Royal Road, Springfield, Va. 22161.

Pesko, Carolyn. *Solar Directory.* Ann Arbor, Mich.: Ann Arbor Science, 1975.

Roberts, Rex. *Your Engineered House.* New York: M. Evans & Co., Inc., 1964.

Shurcliff, William A. *Solar Heated Buildings: A Brief Survey.* Cambridge, Mass.: 1977.

Shurcliff, William A. *Thermal Shutters and Shades.* William A. Shurcliff, 19 Appleton St., Cambridge, Mass. 02138.

Solar Age Catalog. SolarVision, Inc., 200 East Main St., Port Jervis, N.Y. 12771.

Watson, Donald. *Designing and Building a Solar House.* Charlotte, Vt.: Garden Way Publishing, 1977.

Yanda, Bill and Rick Fisher. *The Food and Heat Producing Solar Greenhouse.* Santa Fe, N.M.: John Muir Press, 1977.